South African Weather

and Atmospheric Phenomena

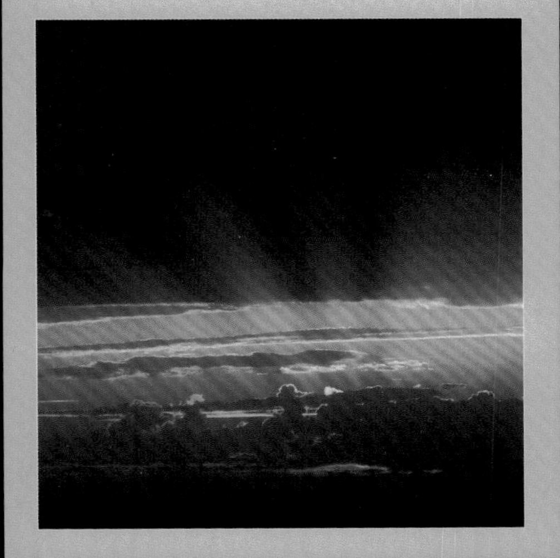

South African Weather

and Atmospheric Phenomena

Dries van Zyl

BRIZA

Title pages: Winter sunrise viewed from the Long Tom Pass, Mpumalanga

These pages: Eastern Free State, to the north of Bethlehem, near sunset

Published by
BRIZA PUBLICATIONS
CK 90/11690/23

PO Box 56569
Arcadia 0007
Pretoria
South Africa

First edition, first impression, 2003

ISBN 1 875093 32 X

Catalogue information: landscape meteorology; weather photography; lightning; atmospheric phenomena; clouds

Managing editor: Reneé Ferreira
Editor: Frances Perryer
Cover design: The Departure Lounge
Maps: CartoCom, Pretoria
Layout: Alicia Arntzen, Lebone Publishing Services
Reproduction: Castle Graphics, Johannesburg

Printed and bound by Creda, Cape Town

Contents

Acknowledgements

My love of mountains and the natural landscape led to hiking trips and landscape photography, which naturally included clouds, storms and other weather phenomena. When I read the book *Rainbows, Halos, and Glories* by Robert Greenler I realised that my casual observations could be transformed. A wealth of weather phenomena was waiting to be discovered and appreciated – if only I would learn to see. I started organising my slide collection, recognising phenomena photographed but not understood or even noticed before, and began concentrating on filling the gaps. As time passed by my book and slide collections grew and so did my awareness of weather phenomena. The authors and publishers of these books, by sharing their knowledge and love of this subject, enable us to know more and therefore see more when weather watching.

In the same way, the World Wide Web has become a gateway to knowledge and sharing. Just search for 'intercloud lightning', for example, and a number of impressive websites will be listed, to be browsed through for free. Eumetsat allows the conditional free use of satellite images on their website. The Web has allowed enthusiastic individuals to self-publish. Strangers throughout South Africa and from the far corners of the globe have stumbled into my website and responded by e-mail, sharing their emotions. An unforeseen kind of visitor was the expatriates who poignantly expressed their longing for the South African landscape they had lost.

The organised photography club scene has competitive leanings, with exhibitions and lectures forming the mainstay of its activities. It is often criticised for inbreeding and staleness, repetition rather than innovation. But it is also a body of diverse, enthusiastic and friendly people sharing a great love for photography, willing to share their knowledge and time. Platforms are created for photographers to exhibit their work and receive feedback, fuelling the quest for excellence. The AFO photography club in Pretoria, founded in 1955, Interfoto, a group of Pretoria clubs organising the bi-annual Pretoria International Exhibition of Photography, and the Photographic Society of South Africa have long been formative forces in my photography.

Ms Elsa de Jager, Manager: Climate Information, Prediction & Publication Services of the South African Weather Service was most helpful in providing information on the SAWS. Together with Ms Colleen Rae, Assistant manager: Training, we had a session on cloud identification during which other bits of useful information also surfaced.

Sometimes, maybe after having gone to bed, I would receive a telephone call urgently calling my attention to something in the sky – three such images included in this book owe their existence to Coen Coetzee and Rip van Wyk. My wife, four sons (while growing up) and extended family members have all endured holiday driving schedules dictated by sunrise and sunset times and weather forecasts, with detours and emergency stops on the road designed for photography.

In 2001 I sent an e-mail to Briza Publications with a proposal for this book. Soon afterwards came the reply: 'Your publication looks quite interesting and Briza will probably be interested in publishing it'! We met in June and our collaboration has been smooth. Ms Reneé Ferreira, the project manager with whom I regularly met, contributed much to the final format into which the book evolved. Ms Frances Perryer was our linguist and queried every vague paragraph.

Introduction

Overview

The book consists of two parts. Part 1 features the subjects that are usually associated with the term 'weather': clouds, storms and rain, cold fronts, climate and weather forecasting. Part 2 is about optical atmospheric phenomena: the dispersion of sunlight into its component rainbow colours by water drops and ice crystals in the atmosphere, which creates a host of spectacular phenomena – some of them common, others rare.

The selection of weather phenomena features those which a keen observer in this region is likely to encounter, over time. They were, likewise, photographed over some time. Weather observation cannot be scheduled much in advance and the observer needs to be ready and prepared when special conditions such as a cold front, a well-placed thunderstorm or a rainbow occur. The sky displays the most predictable events in astronomy, along with the most unpredictable in meteorology.

Cirrus clouds over Table Mountain, as seen from Bloubergstrand

Weather enthusiasts

The term 'landscape meteorology' was defined by the meteorologist Bonacina in 1939, as 'those scenic influences of sky, atmosphere, weather and climate which form part of our natural human environment'. The book emphasises a natural weather experience in this context, rather than the abstract modelling of weather processes.

South Africa has scenic variety and a good climate, punctuated by occasional wild weather created mainly by local thunderstorms or large-scale cold fronts moving into the subcontinent. This combination sets the stage for beautiful landscape meteorology.

Travellers

The images presented here were selected with the emphasis on scenic content and the natural environment of South Africa and feature many of the country's most scenic destinations.

Farmers

Weather and seasonal forecasting and its limitations, the South African Weather Service, the law by which it is governed and its services are discussed. Guidelines for coping with adverse and disastrous weather are included and historically severe weather records are listed. An overview of the South African climate and its major climate regions is given.

Photographers

Landscape photography with an emphasis on meteorology provides a rich spectrum of scenic and colourful subjects. Photography guidelines are included for those who desire to further pursue this somewhat specialised skill, both for personal fulfilment and also to visually communicate special outdoor experiences to others. After all, that is what photography is about: but before technology, it was possible only for painters – notably John Constable. Some thoughts on weather in art are included.

Contact information

The author may be reached via www.weatherscenes.com and visitors are most welcome. This book is not completely comprehensive – has someone out there photographed a tornado, ball lightning or the green flash?

Part 1
Weather

Weather and climate

What we call 'the weather' describes the day-to-day atmospheric conditions of a place. It is experienced as cold and heat, wind, humidity, cloudiness, fog and rain – and whether it is 'good' or 'bad' is largely a matter of opinion. 'Meteorology' is a more objective term, indicating that variables such as temperature, humidity, wind strength and precipitation have been recorded by instruments and measured in physical units. Other parameters such as cloud cover are recorded by humans or instrument platforms such as satellites.

'Climate' is the long-term average seasonal weather of a geographical region, taken over 30 years or more. Weather extremes and frequencies of occurrence are also taken into account. For instance, two different climatic regions may receive the same annual rainfall, but it may rain much more irregularly and in concentrated downpours in the region with a semi-desert climate.

Over time, climate is unstable. The present favourable climatic conditions have led to explosive human population growth, and human activities now influence climate to a debatable extent. This trend may be reversed by human intervention (with mixed success), exhaustion of natural resources (which is a matter of time), or a drastic climate change (unexpectedly).

Sunny weather (and humans) at the gardens of the National Botanical Institute

Stormy weather (and no humans) at the gardens of the National Botanical Institute

The creation of weather

The earth daily spins around its axis (which runs through the poles, almost vertical to the orbital plane around the sun), continually turning the side facing the sun. The equatorial area receives the maximum amount of sunlight, while towards the poles the intensity gradually falls away. Heat is absorbed differently by the atmosphere, land and sea. Clouds and snow-covered land reflect sunlight. Together, these influences cause the earth and atmosphere to be heated in a highly irregular way.

Equilibrium of the temperature of the atmosphere is maintained by the movement of warm equatorial air towards the poles and cold air towards the equator, creating cyclones where they meet (see also 'The climate of South Africa'). As the air moves, it is redirected by the earth's rotation. A complex, ever-changing weather pattern emerges.

Earth's atmosphere is special in that it contains water vapour and has a temperature range that enables water to exist in gaseous, liquid and solid form. Water vapour cannot be sensed by humans.

Previous pages: Typical highveld summer evening, seen from Erasmusrand

Evaporation of moisture from the sea into the atmosphere

The water cycle

Moisture evaporates from oceans, the land surface, lakes and rivers, and is also extracted from the soil and given off by plants. Nearly 90 per cent of the moisture in the atmosphere comes from the oceans. If air cools enough, water vapour condenses into tiny drops of water or ice crystals, forming clouds, fog and dew. This water falls to earth as precipitation, especially over mountain ranges. About 23 per cent of all precipitation falls on the land; the rest falls on the sea. Some water is absorbed by plants and the soil, while the rest finds its way back to the sea, via rivers and underground streams. A water molecule spends about 10 days in the air, and the atmosphere contains about 0,35 per cent of the water on earth.

Because of South Africa's comparatively dry climate, the proportion of rainfall that finds its way back to the sea is much lower than the world average (see also 'The climate of South Africa').

Rain in the Drakensberg foothills feeding rivers

The Blyde River flowing to the sea

A mature thundercloud with anvil top evolved into a Cirrus cloud spreading out on reaching the tropopause

The atmosphere

The layers of the atmosphere

Earth's atmosphere extends into space in a series of layers defined by temperature characteristics. Weather and life occur in the thinnest and lowest of these, the troposphere, whose lower reaches are free of the temperature extremes and deadly radiation that are found higher up. The troposphere reaches up to 16 km above the surface of the earth over the tropics and 8–10 km over the poles, with some seasonal variation. Temperature decreases steadily with altitude (see 'Temperature lapse rates' below), dropping to about –50° C. The troposphere contains more than 80 per cent of the air in the atmosphere.

At the top of the troposphere the temperature profile changes from a lapse condition (temperature decrease with height) to an isothermal condition (temperature constant with height). This boundary between the troposphere and the second layer, the stratosphere, is called the tropopause and is in effect a ceiling on the world's weather.

In the lower stratosphere temperatures do not vary significantly with altitude, but then begin to increase with height, due to energy absorption by the concentration of ozone in this layer. This temperature inversion (temperature increase with height) suppresses vertical air movement and little vertical mixing occurs above the tropopause.

A sensitive balance is maintained between oxygen atoms, oxygen molecules (containing two oxygen atoms) and ozone molecules (containing three oxygen atoms) in the stratosphere. Ultraviolet light radiated by the sun is absorbed by ozone, which generally prevents dangerous radiation levels at the surface of the earth. Natural ozone holes exist above both poles, but the release of certain gases into the atmosphere by humans may destroy ozone and lead to larger ozone holes, allowing dangerous levels of radiation to pass.

Nacreous clouds form in the low stratosphere, at heights of 20–30 km. Resembling thin Cirrus clouds with strong irisation (see 'Atmospheric phenomena'), they remain bright as the sun sets, while lower tropospheric Cirrus grows dark.

Beyond lie the mesosphere, the thermosphere (or ionosphere) and the exosphere. The rare noctilucent (night-shining) clouds occur near the mesospause, about 85 km high. They are mainly seen at high latitudes during summer, against the night sky long after twilight. They are possibly composed of ice that has condensed on meteoritic dust particles. The source of the water vapour is now thought to be small water comets (see 'Comets' below).

The high temperatures of the thermosphere (2 000° C) offer some protection against meteors and space debris of human origin, causing them to burn out as they enter the atmosphere. Ions and electrons occurring here make long-range radio broadcasts possible, by reflecting radio transmissions back to earth.

The troposphere contains clouds from Stratus near the ground to Cirrus at about 6 000 metres, as seen here in the Western Cape

The distribution of clouds in the atmosphere

Clouds are basically classified according to height and form, whether stratiform or cumuloform (see 'Cloud catalogue'). Height determines whether water is present in water or ice form. The exceptions to this system are Cumulonimbus or thunderclouds (see 'Thunderstorms and lightning'), which extend from low altitudes up to the stratosphere.

Atmospheric pressure

Humans are generally unaware of atmospheric pressure, though we are highly adapted to it. Mountaineers and divers know their lives depend on correct management of their exposure to pressure changes. About half of the atmosphere lies below 5 km, and 95 per cent below 30 km. At a height of 3 000 metres, the average height of the Drakensberg, one-third more air needs to be inhaled to obtain the same amount of oxygen normally inhaled at sea level. At the altitude of Mt Everest at 8 858 metres, less than a third of sea-level oxygen is available. (See 'Humans and weather hazards'.)

Earth is surrounded by a relatively thin layer of air, which consists of countless molecules in motion. This motion causes air pressure to be exerted in all directions. Molecules have mass and are pulled towards Earth by gravity, creating atmospheric pressure. Air may be compressed and therefore molecules accumulate near the surface.

Atmospheric pressure is highest at sea level and decreases with increasing height above sea level. The altitude of a location may therefore be deduced from atmospheric pressure, with some reservations. Atmospheric pressure does not decrease with altitude at a regular rate, but drops by about one-thirtieth per 275 metre rise in altitude.

Humid air is lighter than dry air and the water vapour content of air lowers air pressure. For a given altitude, a drop in atmospheric pressure therefore indicates the presence of moisture and the likelihood of rain. Atmospheric pressure also varies horizontally because of the preferential warming of air by land. Warm air is less dense and has a lower air pressure.

Measurement of atmospheric pressure

Atmospheric pressure is defined as the weight of a column of air of unit cross-section. This definition incorporates the force of gravity. Atmospheric pressure at sea level equals a weight of 1 033,6 grams per square centimetre under normal conditions. Normal conditions are considered to occur at 45° latitude at a temperature of 0° C.

Pressure is also defined as the force acting on a unit area. This definition is independent of gravity. A force of one newton per square metre is defined as one pascal. In meteorology the pressure unit hectopascal (hPa) is used internationally. (A hectopascal is 100 pascals and equivalent to the millibar formerly used.) Atmospheric pressure at sea level is 1 013,2 hPa.

An isobar is a line on a weather chart showing equal atmospheric pressure, regardless of the units it is measured in.

Snow on the Amphitheatre, Drakensberg

Temperature lapse rates

Hikers camping on the contour path at 2 000 metres in the Drakensberg may experience a minimum temperature of –6° C in winter, while those camping on the escarpment at 3 000 metres may experience –12° C.

Decreases in temperature and atmospheric pressure with altitude are major parameters in weather processes. The temperature lapse rate is the fall in temperature with increasing height in the atmosphere. An adiabatic process is one in which no heat enters or leaves the system (e.g. a parcel of air). When air is forced upwards, by topography or converging winds, it expands and cools adiabatically at a certain lapse rate.

The dry adiabatic lapse rate is the rate at which a parcel of rising air cools, or sinking air warms with height, without condensation taking place. The dry adiabatic lapse rate is almost 10° C per 1 000 metres.

The wet (or saturated) adiabatic lapse rate is the rate at which a parcel of saturated rising air cools with height, with condensation taking place. Condensation releases latent heat that retards cooling of the rising air. The wet adiabatic lapse rate depends on altitude and differs most markedly from the dry adiabatic lapse rate near the surface of the earth.

The average adiabatic lapse rate is about 6,5° C per 1 000 metres. Rising air will cool at the dry adiabatic lapse rate until saturation followed by condensation occurs. Thereafter it will cool wet adiabatically.

Extraterrestrial influences

The solar constant

The sun is the source of energy driving the weather on earth. The solar energy falling onto the earth in one day is 100 times that of a strong earthquake, 10 000 times that of a tropical cyclone, 100 000 times that of a nuclear explosion, 100 million times that of a summer thunderstorm. Above the atmosphere, about 1 370 watts per square metre of solar energy flux fall on earth, at its mean distance from the sun, a figure known as the solar constant.

The solar constant slowly increases over millions of years as the sun evolves and becomes hotter. In shorter time scales it changes with the solar cycle (see below) and sunspot activity.

At the time of the first dinosaurs, about 300 million years ago, the solar constant was about 2,5 per cent less than today. As the earth's climate evolved, suitable conditions developed for life forms to flourish notwithstanding a changing solar constant, and this strongly supports the Gaia hypothesis of the earth as one living organism (see also 'The climate of South Africa').

The radiation balance

The earth as a whole maintains an energy balance. Energy is absorbed and radiated back into space by the earth, and absorbed, reflected and scattered by the atmosphere. About 51 per cent of sunlight penetrates the atmosphere, while 6 per cent of the earth's radiation escapes into space. The energy absorbed by the atmosphere fuels the weather.

The so-called greenhouse effect exerted by the atmosphere is due to selective absorption of radiation by clouds. Heat absorbed by the earth and radiated into space throughout the night is absorbed at the base of low clouds and radiated back, while other radiation passes through. Without the greenhouse effect the earth would be too cold to support life. But an increased greenhouse effect caused by pollution may disturb the radiation balance of the earth. (In a true greenhouse convective heat loss is also prevented.)

During the day more heat is absorbed than lost and the temperature of the surface of the earth rises, reaching its maximum a few hours after solar noon. At night the surface cools down and the minimum occurs just after sunrise.

Variations in the solar constant (see above) influence climate in ways largely unanswered. In the long run the earth's climate is erratic, as revealed by research into earlier climates.

The sun

Earthshine, sunlight reflected by the earth and falling on the moon, is seen in the dark part of the moon, while the brilliant crescent and Venus are lit directly by the sun, and the star seen faintly above to the right is radiating light by itself.

Seasons

The earth annually orbits the sun, moving in closest on 2–3 January during the southern summer and farthest out on 5–6 July during the northern summer. The earth's axis is tilted about 23,5° towards the orbital plane, causing the southern and northern hemispheres to alternately face the sun more directly, creating the seasons. Seasons have a marked influence on climate and weather. Changes in the earth's orbital parameters, over thousands of years, have a significant influence on seasons (see also'The climate of South Africa').

Glen Reenen, Golden Gate National Park, in spring

Glen Reenen, Golden Gate National Park, in autumn

The solar cycle

The occurrence of small changes of about 0,1 per cent in solar output (see 'The solar constant' above) over an average 11-year cycle is known as the solar cycle. The cycle varies from 9 to 13 years and has some influence on weather. The 11-year cycle in the number of sunspots visible on the sun, and the orbital period of Jupiter, the largest planet, coincide with the solar cycle, and they are thought to be related.

The planets

The alignment of the planets is a popular traditional scarecrow, though in fact it is a fairly regular phenomenon and passes unnoticed by most. The combined gravitational pull of all the planets has a minor effect on the sun and its radiation output (see also 'The solar constant' above).

The moon, Venus and Jupiter aligned and seen in daylight on 23 April 1998, with Venus about to be occulted (hidden) by the moon

The moon

The moon and its phases do not influence the weather in a detectable way (see also 'Personal fore-casting' in 'Forecasting'), but the moon does have a cyclical influence on climate.

The gravitational pull of the moon is about one-sixth that of the earth. It varies in accordance with the moon's distance on its elliptical orbit around the earth. It supplements or opposes the gravitational pull of the sun to varying degrees, creating ocean tides and also less-pronounced tides in the atmosphere and the crust of the earth. Ocean tides have little or no influence on the weather. But ocean tides had a marked influence on the moon, arresting its rotational speed so that the same side always faces earth.

The strongest tidal forces are created when the earth, moon and sun are optimally aligned and closest, during 90-year and other, longer cycles. These forces increase vertical mixing of water in the oceans and bring cold water to the surface, which influences climate.

Solar eclipses

During a solar eclipse a small part of the earth is shielded from the sun, causing temperatures to drop. But it is of too short a duration, with totality lasting only a few minutes, to impact on the weather. Weather often obscures the sun, leaving eclipse chasers disappointed.

Comets

A comet's nucleus contains frozen water and a mixture of frozen gases. The controversial new Small Comet theory purports that small (house-sized) water comets crash into the

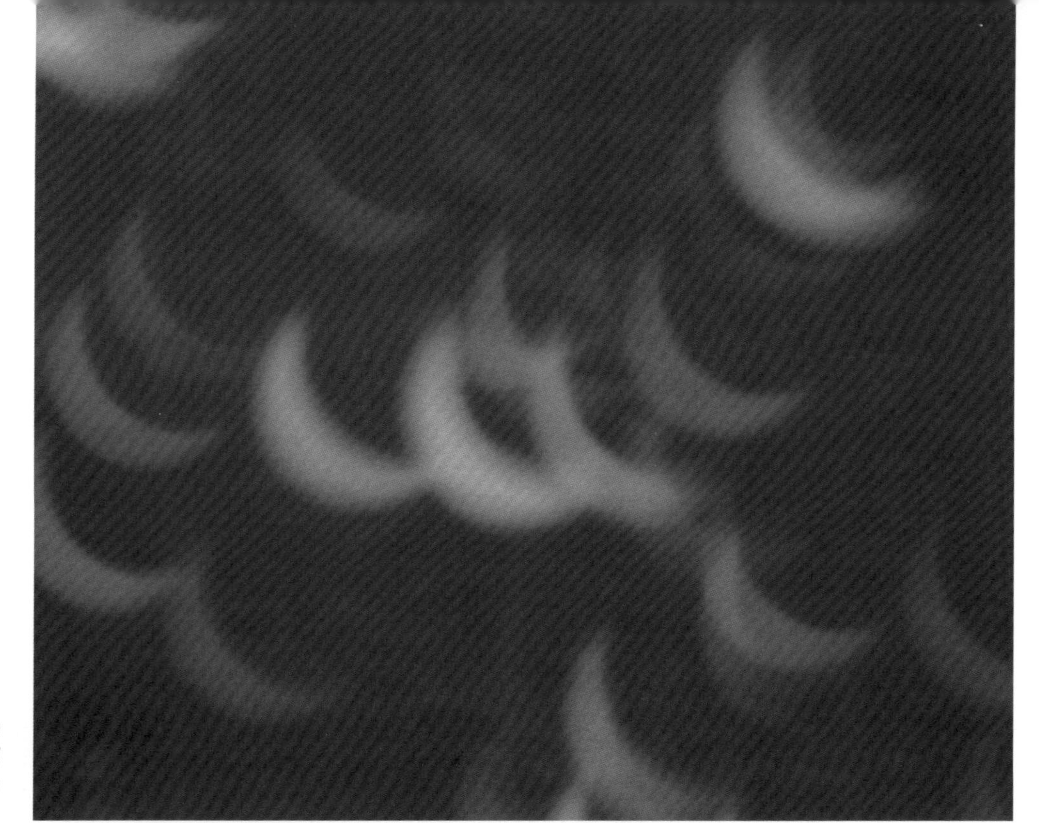

Light from the partially eclipsed sun is focused through tiny gaps between leaves.

Comet Hale-Bopp in the sky above Pretoria in March 1997

earth's atmosphere at a rate of about 40 per minute. They add to its total moisture content – and may indeed be the source of all water on earth. It is estimated that small comets add approximately 2,5 cm of water to the surface of the earth every 20 000 years. They may be the source of the high-altitude noctilucent ice clouds (see also 'The layers of the atmosphere' above).

Meteors

Most meteors burn out in the atmosphere, leaving behind dust particles. The Hoba meteorite in Namibia is the largest meteorite found on earth. A much bigger meteorite hit the earth at Tswaing north of Pretoria about 120 000 years ago and must have blasted huge amounts of soil into the atmosphere. The impact occurred fairly recently in geological times, which explains why the impact crater is so well preserved, one of the best in the world.

The Hoba meteorite in Namibia

The Tswaing meteorite impact crater to the north of Pretoria

Above: Winelands near Stellenbosch

Coping with adverse weather

Farmers

Farmers are aware of daily, seasonal and annual variations in the weather – for example, the occurrence of early and late frost that determines maturity dates. Unpredictable weather and unpredictable markets are the farmer's main adversaries. The discipline of agrometeorology strives to support crop production through the study and use of weather and climate information. Seasonal forecasts, also made by the SAWS (see also 'Weather forecasting'), will hopefully improve markedly in future.

Weather insurance

The exposure of farmers to certain weather hazards can be evaluated statistically, which makes insurance a viable prospect for insurance companies. Insurance, if available, is the only protection against many hazards, at a price. The cost of the insurance gives the farmer some indication of the magnitude of the risk he carries, beyond what he already knows.

Weather derivatives

The term 'weather derivatives' is used for risk management contracts available to industries such as agriculture whose business is affected by the weather. Weather derivatives are based on the principle that the risk of exposure to a weather hazard for one industry is balanced by investment in another industry that stands to profit from the same conditions. For example, farming industry losses to above-normal seasonal heat are offset by manufacturing industry gains in orders for cooling equipment.

Consulting services, training courses presented by South African universities and software tools are available options.

Droughts and flooding

Droughts and floods are very much a part of South African weather. Crop farming dependent on rain will, sooner or later, fail for a season – and farmers have somehow to bridge drought years.

Drought is a relative concept and there is no universal definition. Various definitions are tailored for specific farming or wildlife management activities in different climate regions. Timing of rainfall is a crucial element in crop production, where certain growth phases are more sensitive and the monthly rainfall average is meaningless then.

Level farmlands are susceptible to broad-scale flooding, for example maize farms in the Free State. Farmlands on inclines are susceptible to flash floods – the contouring of lands provides protection. Farmlands bordering on rivers are vulnerable to flooded rivers. Seasonal forecasts endeavour to evaluate the risks and a farmer may decide to choose another option if the forecast is bad for any specific one.

Storms

Hail storms do not last long and follow a narrow band (see 'Thunderstorms and lightning'). A sizeable area of crop is therefore at less risk as a whole. Hail storms are predictable statistically in the context of climate and season, but unpredictable in terms of date and locality. The only feasible protection farmers can have against hail is insurance.

Lightning rods safeguard buildings against lightning damage and fire caused by lightning (see 'Thunderstorms and lightning'). The low incidence of lightning does not justify the safeguarding of general livestock.

Strong winds may blow away valuable topsoil (see also 'Dust storms', under 'Wind storms'), eventually rendering the lands unusable. They may uproot plants and damage buildings.

Cold and dry winds cause black frost. The dewpoint is not reached above freezing point, which would have allowed the formation of a protective layer of frost. The fluid inside cells freezes and the resulting expansion kills off the cells. Strong warm winds will wither young plants.

Hail damage

Severe damage is sometimes caused by microbursts or downbursts (see also 'Microbursts' under 'Wind storms'), misleading observers after the event to believe it was caused by a tornado. Damage to Rosendal in the Free State on 20 April 1994, and Sodwana Bay on 22 April 1995 were wrongly ascribed to tornadoes by the press.

There is little a farmer can do when hit by a tornado besides counting the damage, should he survive. A database of South African tornado activity is maintained by the

CSIR. By mid-1996 it contained about 200 tornado events since the first on 3 December 1905. Since 1948 data collection has become more reliable – this period contains 174 tornadoes. The eastern part of the country is more prone to tornadoes. The peak period is mid-summer and the peak time is between 16:00 and 19:00. On average four tornado events occur per year. Almost 80 per cent travelled in a general eastward direction. Path lengths are available for some 31 tornadoes and vary from 5 to 170 km with an average of 27 km. Path widths vary between 20 and 300 metres, although the path of the 1990 Welkom tornado was 1 700 m wide.

Heat waves

A heat wave exists when the maximum temperature exceeds the mean maximum for the hottest month for a period of three days. A heat wave during a period of drought may be the last straw for crops in a vulnerable growth stage. Heat waves also increase the risk of fire.

Anticyclonic conditions, high pressure systems with descending air and atmospheric stability, are a dominant feature of southern African weather and climate. In combination with warm and dry berg winds they produce heat waves in coastal and adjacent areas. Temperature increases exceeding 5° C from one day to the next occur about 15 times annually in the interior north of Port Elizabeth, about five times in the southern half of the country and perhaps once in the northern half.

Cold snaps

Cold fronts should be seen as an integral part of climate and farming planned accordingly. Planting crops susceptible to frost in early autumn or late spring is risky. Expensive glass houses provide protection and extend the season. Only severe cold fronts bring snow in quantities that pose a threat to livestock, and usually only in the regions adjoining the Drakensberg.

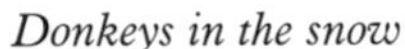

Donkeys in the snow

Humans

Cold

Marked variations in temperature are unhealthy. Cold weather increases blood pressure and reduces resistance to infections. Humans are able to endure amazingly low temperatures, even when naked, but the danger arises when cold combines with wind to produce wind-chill.

Wind-chill

Exposure to wind causes body heat to be lost more quickly, through evaporation of skin moisture and removal of warm air next to the body. In combination with wet clothes, low temperature and fatigue, it leads to hypothermia. This is a far greater hazard than lightning, for example, and over the years a number of hikers have perished in the Drakensberg and Western Cape mountains. Wind chill has been called 'the killer of the unprepared', but it can be offset when equipped properly. There is a saying that there is no bad weather, only bad clothes.

A wind-chill table shows the apparent temperatures produced by a combination of physical temperature and wind speed, as they are experienced by humans. For example, a temperature just below zero in combination with a strong wind will effectively be equivalent to a temperature of about –20° C in still conditions. Clearly, a person trapped outdoors should seek shelter against the wind. The coldest temperatures tabled may be reached in South Africa on high mountains and elsewhere during cold snaps.

Wind-chill table					
Wind speed (kph)	Actual temperature (°C)				
	4	–1	–7	–12	–18
	Apparent temperature				
24	–5	–13	–21	–28	–35
32	–7	–16	–23	–31	–39
40	–9	–17	–26	–34	–42
48	–11	–19	–28	–36	–45

Heat

Sustained high temperatures known as heat waves do not present a major threat to humans in South Africa, as in some other countries such as India where many people sometimes perish in temperatures way above 40° C. Indirect threats in the form of tropical diseases may increase.

Lightning

Lightning is a deadly but statistically small weather hazard. Although the Drakensberg is a high lightning incidence region and a popular outdoor recreational area, only two fatal incidents involving hikers have been recorded, and in both cases basic safety precautions were neglected. (Visitors to the Golden Gate National Park may visit the graveyard where those killed long ago near Mont-aux-Sources are buried.) About 10–20 people are killed annually in South Africa, but figures are probably under-reported. The hazards of lightning are described with personal safety guidelines under 'Thunderstorms and lightning'.

Flooding

Flash floods pose a major threat to humans visiting canyons and gorges, especially when it rains higher up in the catchment area and the flood arrives without warning. Necessary safety precautions include official warning and evacuation protocols or communication with the outside world. Camping in dry river beds or next to streams is risky. Hikers have been killed in the Drakensberg and on occasion have had to evacuate camp on the Fish River hiking trail in Namibia. Crossing swollen streams is highly dangerous and the prepared hiker will be able to wait for as long as it takes for the waters to subside. (Also see 'Floods' under 'Precipitation'.)

Tornadoes

Tornadoes generate the strongest winds. Humans are pretty defenceless when caught out in the open, in passenger cars or homes. Cars may be blown over and the roofs of homes sucked up and deposited elsewhere. Flying debris poses grave dangers; sheltering against a wall provides some safeguard, unless the wall is blown over. In the worst incident in South Africa, 24 people were killed and more than 600 injured in Albertynsville on 30 November 1952. One death occurs per two tornado events, on average, leading to about two deaths per year, but this figure is probably higher due to under-reporting.

Altitude

About half the atmosphere lies below 5 km. At a height of 3 km, the average height of the Drakensberg, one third more air needs to be inhaled to obtain the same amount of oxygen normally inhaled at sea level. (See 'Weather' for a discussion of atmospheric pressure.) Altitude sickness resulting from oxygen starvation (hypoxia) presents a major hazard in the world's high mountains. Although the Drakensberg is much lower, people have developed symptoms of this affliction and died at this height elsewhere. The danger lies in rapid height gains, for example by driving from the coast to a height of 2 000 metres and climbing to 3 000 metres on the same day. The potentially fatal process is reversed by an immediate descent. If South Africans climbing Kilimanjaro would only spend a couple of days acclimatising, for example by first climbing Meru, they should scale the higher mountain without the usual headaches, nausea and disturbed sleep patterns.

Humans become acclimatised to altitude, up to a limit. The highest altitudes permanently inhabited are about 5 100 metres, but the lack of oxygen impairs the physiology of the human body. Intellect and physical performance are adversely affected, at first subtly, from as low as 1 500 metres, according to the authority Charles Houston (see Bibliography).

Drakensberg escarpment at Injasuti

Biodiversity in hot and dry climates

South Africa is a country with predominantly hot and dry climate regions. In the Kalahari the ground surface can be as hot as 70° C, while a temperature of 40° C is measured inside a Stevenson Screen (see also 'Weather forecasting'). A couple of examples of a rich diversity of highly adapted life forms follow.

Mammals Animals need drinking water to replenish water lost through evaporation of sweat, the mechanism used to maintain normal body temperature. Death usually occurs if brain temperature rises to 42° C. Gemsbok, however, do not need drinking water. They possess a remarkable temperature regulation mechanism that keeps the brain cool when blood temperature rises on very hot days in the Kalahari. The

Gemsbok in the Kgalagadi Transfrontier Park

body temperature is allowed to rise to 45° C and blood is cooled by about 3° C immediately before it enters the brain. The cooling is effected by an arterial network utilising evaporation of moisture in the nose. The closely related springbok is suspected to have a similar mechanism.

Birds The doublebanded courser is adapted to the erratic Kalahari environment. It does not need drinking water and can withstand extreme temperatures. It breeds three to four times throughout the year, whatever the conditions, laying only one egg. At temperatures below 30° C the egg is incubated to warm it, between 30° and 36° C it is simply shaded, and above 36° C it is incubated to conduct heat away.

Trees The majestic camelthorn tree of the Kalahari initially grows very slowly above the ground while developing an extensive root system, until groundwater is reached.

Namaqualand's wild flowers in spring

Flowers Namaqualand's wild flowers compensate for hot, dry summers by completing the life cycle during spring, the seeds then lying dormant until the next spring.

The recently discovered *Clivia mirabilis* of Oorlogskloof, near Nieuwoudtville, flourished during times when the climate there was more subtropical, as do present-day plants 800 km away in shaded and moist spots in the eastern part of the country. But as the climate in the west of the country became hotter and drier during summer, the *Clivia mirabilis* became isolated and adapted.

Extreme past weather

Some South African extreme weather records are presented in the following table. These were recorded at official weather stations and therefore have probably been exceeded elsewhere, especially at altitude. For example, winter night temperatures of –15° C are typical in the Drakensberg at altitudes above 3 000 metres; there the lowest temperature record of –18,6° C is probably exceeded frequently.

Record weather			
Subject	*Magnitude*	*Locality*	*Date*
Highest temperature	50,0° C	Sundays River Valley	3 November 1918
Lowest temperature	–18,6° C	Buffelsfontein, near Molteno	28 June 1996
Highest rainfall in 24 hours	597 mm	St Lucia Lake	31 January 1984
Highest rainfall in a calendar month	1 845 mm	Entabeni, Limpopo Province	January 1958
Highest rainfall over one year	3 874 mm	Jonkershoek	1950
Highest wind gust	186 kph	Beaufort West	16 May 1984
Longest tornado path	170 km	Hanover/Trompsburg	1976
Widest tornado path	1 700 m	Welkom	1990

Rainfall, temperature and wind are generally the prominent weather parameters for South Africans. Many more parameters may be identified, such as driest location, greatest temperature change in one day, most rapid temperature change, coldest and hottest annual average, highest and lowest air pressure, largest hailstone, windiest location.

Weather forecasting

Weather forecasting improved greatly when synoptic weather data became available. The first milestone was the invention of the telegraph in the 1830s, which made it possible to transmit weather data from different weather stations without delay to the forecasting centre. The second major advance came with the launch of weather satellites – such as Meteosat, which serves South Africa. Advances are continually being made with successive generations of supercomputers used in numerical weather prediction.

Mathematical weather models simulate the motion of the atmosphere and the physical processes within. The models are solved numerically. In order to do this the atmosphere is divided into discrete volume units with uniform properties that interact. Volume units have, say, 80 km sides and are stacked 20 units high in the atmosphere. Smaller units provide greater accuracy and demand more computing power, driving the demand for ever-more-powerful computers.

The computer models predict future weather patterns using formulas that describe the dynamic atmospheric processes only approximately. Human forecasters prepare forecasts that may combine this information with satellite and radar images that arrive

Above: Crepuscular rays at sunset, Erasmusrand

after the deadline for the computer analysis. They may take local topography into account. They use their experience and judgement and the final forecast is a mixture of art and science.

Seasonal forecasting predicts above-average or below-average seasonal rainfall and temperature – and here the El Niño phenomenon (see also 'The climate of South Africa') has become prominent.

Weather maps

Weather maps show either prevailing or predicted weather. A synoptic chart shows the weather over a geographic region at a certain time. It starts off as a simplified map showing only the most important features such as coastlines, national and provincial borders, large rivers and cities. Lines are drawn between points of equal air pressure, forming isobars indicating the spatial distribution of areas of equal air pressure. Closed isobars indicate areas of high and low pressure, labelled H and L. With highs, air pressure increases towards the centre; with lows, it decreases towards the centre. Isobars may be marked with the air pressure expressed in hectopascals (see also 'Weather'). Cold and warm fronts, the boundaries between air masses (see also 'The climate of South Africa'), are marked on the map (usually from satellite images). The position of a cold front is indicated by blue triangles, while red semicircles are used for a warm front. Alternating symbols indicate an occlusion, where two air masses overlap. Warm fronts are rarely seen over land on South African weather maps.

Satellite image with a cold front approaching Cape Town

Corresponding synoptic chart

Reading a weather map

Isobars reveal air pressure patterns and from them the direction of weather systems may be deduced. Air moves from high- to low-pressure regions across isobars but is deflected to the left (in the southern hemisphere) by the Coriolis effect, which causes winds to blow towards and clockwise around lows, and away from and counter-clockwise around highs. Isobars may be followed out of the highs and lows into the regions where the winds come from, revealing the kind of winds to be expected, whether moist or dry, warm or cold. Closer spacing of isobars indicates stronger winds and unsettled weather and usually occurs with lows. Isobars further apart usually occur with highs and indicate settled weather and clear skies. Internationally standardised weather symbols are placed on maps to provide more information.

The following table contains a selection from the set of standardised international weather symbols.

International Weather Symbols

Current weather
light drizzle
steady, light drizzle
intermittent, moderate drizzle
steady, moderate drizzle
intermittent, heavy drizzle
steady, heavy drizzle
light rain
steady, light rain
intermittent, moderate rain
steady, moderate rain
intermittent, heavy rain
steady, heavy rain
light snow
steady, light snow
intermittent, moderate snow
steady, moderate snow
intermittent, heavy snow
steady, heavy snow
hail
freezing rain
smoke
tornado
dust devils
dust storms
fog
thunderstorm
lightning
hurricane

Sky coverage
no clouds
one-tenth covered
two- to three-tenths covered
four-tenths covered
half covered
six-tenths covered
seven- to eight-tenths covered
nine-tenths covered
completely overcast
sky obscured
Low Clouds
Stratus
Stratocumulus
Cumulus
Cumulus congestus
Cumulonimbus calvus
Cumulonimbus with anvil
Middle Clouds
Altostratus
Altocumulus
Altocumulus castellanus
High Clouds
Cirrus
Cirrostratus
Cirrocumulus

Wind speed (kph)
calm
1–3
4–13
14–23
24–33
34–40
89–97
192–198

Weather forecasting terminology

The following terminology of subjective weather parameters was developed for public consumption, to be used instead of physical units.

Temperature		
Subjective term	**Physical units (°C)**	
	Summer Oct–March	**Winter April–Sept**
Very hot	>35	>35
Hot	30–34	27-34
Warm	26–29	23-26
Cool	20–25	
Mild		17–22
Cold	15–19	12–16
Very cold	<15	<12

The subjective terms for temperature take the season into account. For example, a temperature of 25° C feels cool in summer and warm in winter.

Wind	
Descriptive term	**Physical units (kph)**
Light	1–16
Moderate	17–24
Fresh	25–35
Strong	36–56
Near gale	57–63
Gale	64–74
Strong gale	75–87

See the chapter 'Storm winds' for the terminology and tables applicable to tornadoes and tropical cyclones.

Cloudiness	
Descriptive term	**Cloud coverage**
Fine	Little or no cloud
Partly cloudy	Covering half or less of sky
Cloudy	More than half but some blue sky visible
Overcast	Completely covering the sky

Only four broad categories of cloud coverage are used.

The sea

'Total sea' (or 'Total wave height') is the average wave height as obtained from wave-measuring buoys. It consists of swell wave and wind wave components. Swell waves are wind waves that were generated some distance away. Much of the swell at the coast of South Africa has been generated more than 1 000 km to the south-west several days earlier, and therefore a heavy swell is possible under windless conditions. Wind waves are generated by the local wind and die out quickly when the wind ceases to blow, with the waves travelling away as swell.

The South African Weather Service

History

Modern meteorology was born in the wake of a weather disaster in 1854, when a French warship and 38 merchant vessels sank off the Crimean port of Balaklava during a storm. The director of the Paris Observatory, Liverier, discovered that the vessels could have been forewarned: meteorological records revealed that the storm had formed two days earlier and had crossed Europe in a south-easterly direction. A national storm-warning system was established, using the telegraph to facilitate data transmission. The first weather maps in modern format were published in France in 1863, and by 1872 Britain was doing the same. Storms on the Great Lakes in North America sank or damaged some 3 000 vessels during 1868-69 and many lives were lost. This led to the establishment of the United States' first official weather service in 1870. In 1878 the International Meteorological Organisation (IOM) was established; in 1951 it became the World Meteorological Organisation (WMO).

Stormy sea at the Cape of Storms

With the promulgation of Government Notice No 363 of 1860, South Africa became one of the first countries in the world to establish a national weather service. The country joined the modern era in 1905, when some 30 weather stations started submitting regular weather reports by telegraph or telephone, on an exchange basis to the Meteorological Commission in Cape Town, the Observatory in Durban, the Meteorological Observer Bloemfontein and Salisbury. These institutions cooperated until the Union Weather Service was established in Pretoria in 1912, under the Department of Irrigation. The first national forecast was issued from the Cape Town branch office on Monday, 15 April 1912. In June 1926 the forecasting service was transferred to Pretoria.

During the Second World War civilian weather services were interrupted in order to serve the Air Force and it took many years to restore normal services after the war. In 1949 the name of the weather service was changed to the Weather Bureau and it functioned under the Department of Transport. During the 1980s the Weather Bureau was transferred to the Department of Environmental Affairs and operated as such until 2001.

Today

On 15 July 2001 the Weather Bureau became the South African Weather Service, operating under the South African Weather Service Act, Act No 8 of 2001. Schedules 1 and 2 stipulate the public good services and commercial services to be provided to the country. The SAWS employs about 350 people, about half of whom are deployed at the Pretoria headquarters and the others at weather offices around the country. It renders services to one tenth of the globe, including the oceans surrounding southern African and neighbouring countries. It cooperates internationally and is a member of the WMO.

Public good services

The thrust of public good services is towards:

- weather and climate forecasting including meteorological data gathering and international cooperation through the WMO;
- research, training, maintenance of the National Meteorological Library and custody of the National Climatological Databank;
- provision of daily rainfall and maximum and minimum temperatures to the general public; aviation and maritime meteorological support; subsistence farming and fishing support;
- warning services intended for the general benefit of the population and safety of life and property.

Commercial services

A selection of commercial services is:

- specialised weather forecasting and climate information services, particularly to the media;
- dissemination of weather and climate information;
- manufacturing, selling and servicing of meteorological equipment;
- consultations including advice to the legal and insurance industries.

South African pioneer of meteorology

Jan Taljaard joined the South African Meteorological Office in 1939, having received a master's degree in physics from the University of Stellenbosch the previous year. He later obtained a Ph.D (Meteorology) from the University of the Witwatersrand. His meteorological work and research continued after his retirement from the post of Director of the South African Weather Bureau in 1978. At age 82, he co-authored a chapter in a publication on the meteorology of the southern hemisphere. In collaboration with others he prepared the atlas *Climate of the Upper Atmosphere: Southern Hemisphere*, first published by the United States Navy in 1969. The foundations of much of the meteorological research work conducted in South Africa can be traced back to him.

Meteorological data collection

As a member of the WMO, the South African Weather Service participates in the World Weather Watch programme of international cooperation and regular data exchange. The SAWS has its own vast data collection network, consisting of ground stations with a variety of instruments and devices, and unmanned instrument platforms. Information is gathered from the land and sea surface, from below the sea surface, throughout the height of the atmosphere and from space.

Ground stations

A number of ground stations are spread out throughout South Africa and on Marion Island and Antarctica. About 1 800 rain stations regularly report rainfall to the SAWS. Ground stations have always played a major role in meteorological data collection.

Radiosondes

A radiosonde is a balloon carrying instruments such as a thermometer, a barometer and a hygrometer, and a radar reflector. As it rises through the atmosphere it gathers important data on the vertical distribution of weather conditions and radios it back to earth. Radiosondes are released into the atmosphere from weather ground stations on a regular basis.

Manned balloons played a major and dramatic role in the history of atmospheric research, before unmanned balloons took over in modern times.

Sea buoys

Until recently weather forecasting focused mainly on the atmosphere. It is now recognised that the way energy is stored in the oceans and transferred to the atmosphere is important, especially since more than 70 per cent of the surface of the globe is covered by ocean.

Sea buoys are positioned at regular grid points in the oceans and gather meteorological data at and below the sea surface. The International Drifting Buoy Project has been a major boon to Southern Hemisphere meteorology. South Africa's participation was promoted by Jan Taljaard (see box), among others.

Weather satellites

Meteosat is one of a system of geostationary satellites spaced around the earth at a height of almost 36 000 km above the equator. Their rotation is synchronised with that of the earth and Meteosat therefore always looks down on Africa. It photographs the whole globe and the part covering South Africa usually seen in the media is extracted from the bigger image. Weather systems are viewed from above and the existence and dynamics of large-scale systems are dramatically visible. Other satellites are polar-orbiting and at lower altitudes and record strips of the earth below in more detail. At the Hartebeesthoek tracking station, visual Meteosat images are continuously received throughout the day and infrared images throughout the day and night.

National forecasts

National forecasts are produced at the SAWS headquarters in Pretoria, where all data are collected and prepared as input to numerical weather models running on a supercomputer.

Various forecasts, differing in the selection and level of information, are targeted to specific users. These range from prime-time TV forecasts to regular radio forecasts and daily newspaper summaries. Both short-term weather forecasts and seasonal outlooks are published.

The SAWS maintains a website that features many of its services and products, as well as popular information and links. Weather forecasting is prominent, naturally, and the website attracts many visitors.

Services of interest to farmers

Seasonal outlooks for the next three months are published monthly. They focus on rainfall and temperature expectations and are indicated on a map of the country subdivided into a few broadly defined regions.

'Normal' rainfall and temperature are defined in terms of monthly averages for a particular region and not the seasonal average. Probability forecasts are given, indicating the probability, as a percentage, that weather will be above, near or below normal. For example, a rainfall probability forecast of 50/30/20 indicates that there is a 50 per cent chance that rainfall will be above normal, 30 per cent that it will be near normal, and 20 per cent that it will be below normal. A similar probability forecast is given for temperature. Above-normal rainfall may still be badly distributed spatially and in time.

These forecasts, as well as an 8–14 day outlook and a monthly forecast, are available on the SAWS website http://www.weathersa.co.za/Forecasts/forecast.htm or by telephone and fax.

Only subsistence farmers are singled out by the South African Weather Service Act, in stipulation 13 of Schedule 1.

While the accuracy of daily forecasts faces fundamental limits (see 'The limits of weather predictability' below), general forecasts of a few months, of most interest to farmers, will probably improve markedly in future.

The limits of weather predictability

In the early 1960s the meteorologist Edward Lorenz of the Massachusetts Institute of Technology investigated systems with non-linear properties. In such a system, defined by a number of parameters, changes in one parameter do not cause the others to change in direct proportion. Using an atmospheric convection model, he showed that it was inherently unpredictable, never returned to its initial state and never repeated. Investigation into this kind of behaviour led to the development of chaos theory, which later found many applications in nature. It indicated that there are fundamental limits to weather forecasting, regardless of the accuracy and quantity of data and the power of computers. Tiny errors in weather models will grow with time and overwhelm the analytical results – perhaps within 10 days of the forecast.

It is recognised today that weather worldwide fluctuates between a relatively stationary state, when it is relatively predictable, and a dynamic state when rapid unpredictable changes can be expected. This problem is addressed by a method known as ensemble forecasting. Computer simulations are run repeatedly but with slightly different initial atmospheric conditions. If the ensemble of forecasts does not diverge in time the forecast is expected to be reasonably accurate; otherwise the weather is too unstable for a reliable forecast to be made.

Although detailed forecasting is limited by chaos theory, there is a brighter future for general forecasts of a few months, based on the middle latitudes' long-lasting weather patterns, which sometimes last for weeks, and their causes.

Personal forecasting

Personal weather forecasting is based on observation and interpretation of signs in nature. It may be enhanced by instruments and even a personal weather station.

A possibility of rain is indicated not by the crescent moon but rather by the pale yellow sky in the west at sunset

The moon

Although it is a popular traditional concept, there is no scientific evidence that the phases of the moon influence the weather in any way (for example by increasing the likelihood of rain). On any given day, local weather conditions throughout the country (and world) vary significantly under the same phase of the moon. Weather is driven by energy radiated by the sun and redistributed by atmospheric processes, and the contribution of the moon in any phase to these is negligible. The phases and orbital movement of the moon are highly regular and repetitive, while weather is highly irregular and complex, and weather systems never repeat, indicating no link between the two.

In middle latitudes, forecasting that tomorrow's weather will be the same as today's will be correct about 70 per cent of the time. This is known as the persistence method. Forecasting that tomorrow's weather will be the same as the average seasonal weather will be right to a similar degree. This is known as the climatological method. These no-skill forecasts ignore actual weather. Forecasts utilising modern skills process enormous amounts of data to calculate and predict changes in the weather. Accuracies of more than 85 per cent for 24-hour forecasts and 80 per cent for five-day forecasts are achieved.

Animals and insects sense and react to certain present weather conditions. They do not predict coming weather and their behaviour is a poor guide to the weather. For example, during a solar eclipse in the middle of the day, birds and animals behave as at sunset, simply reacting to the darker and cooler environment. Plants are sensitive to temperature and plant growth is stimulated even more by heat than by rainfall. The spring flowers of Namaqualand only open as the day warms. Insects are more active in hot conditions. Spiders spin webs more prolifically in dry conditions. Although these behavioural changes may indicate the weather to come, they also may not.

Namaqualand flowers, only opening when the weather warms

Predicting rain

The dominant persistent atmospheric motion is eastward. Winds known as the westerlies are centred over the mid-latitudes from approximately 35° to 65° near the surface of the earth and wider in the upper levels of the troposphere. Therefore if air pressure is falling, say overnight, and clouds gather in the west, rain may be expected. If pressure is steady and stratiform clouds are present, rain is unlikely.

Sunrise over Monk's Cowl, Drakensberg

Some signs in the sky may be interpreted without instruments. At sunset or sunrise, red clouds indicate dry air while pale yellow clouds indicate moist air. Since weather usually moves east, a yellow sunset signals that rain is coming. A red sky at sunset signals the approach of dry air and a low probability of rain the next day. A red sky at dawn indicates that the dry air has passed.

High-level clouds, consisting of ice crystals, often precede a strong depression. A halo around the moon indicates the presence of such clouds and the coming depression that will bring rain. A rainbow in the east (afternoons) shows that the shower is moving away, while a rainbow in the west (mornings) indicates that rain is approaching (see also 'Atmospheric phenomena').

Different cloud species reveal different atmospheric conditions (see also 'Cloud atlas').

Instruments

Weather forecasting begins with careful observations and recordings. A rain gauge is used to record rainfall. A wind vane or sock indicates wind direction, which is measured by means of a compass. An anemometer uses spinning cups to drive a dial indicating wind speed. All these instruments should be installed in the open, away from buildings or trees.

Cloud cover is measured in one-eighth fractions, on a scale from zero for no clouds, to eight for a completely overcast sky (see box on 'International weather symbols'). It is conveniently recorded by means of a mirror with a grid, laid down on the ground, by simply counting the grid squares containing cloud.

Measurement of the temperature, humidity and pressure of the air is more reliable when made in a standardised environment. A Stevenson screen is a ventilated louvered wooden box on legs, painted glossy white, used to host the weather instruments listed below. Instruments are protected from direct sunlight while the interior does not heat up. It is easy to construct (see Bibliography) and should be placed not closer than 10 metres from buildings and on a lawn.

A maximum-minimum thermometer records the highest and lowest temperatures during a period, typically per day. A wet- and dry-bulb hygrometer measures the humidity of the air. A barometer measures air pressure. When installed, it has to be adjusted for the altitude of the weather station. Although mercury barometers are most accurate, aneroid barometers are convenient and therefore popular. They need to be recalibrated regularly.

Advanced instruments may be used. A thermograph plots temperature readings and a barograph plots pressure readings against time. Various electronic devices have become available, monitoring several atmospheric parameters. Meteosat weather images are published on web sites.

Artificial weather modification

In ancient – and less ancient – times weather was seen as a tool in the hands of the gods, used to reward or punish humanity, and therefore meddling with the weather was prohibited. But science is also uncertain about how far it may go and concerned about unforeseen complications (see also 'The Gaia Hypothesis').

Aristotle reasoned that if one weather event always follows another, the first may be the cause of the second. The implication is that if the first can be modified, the second may be controlled. This idea is followed through today, by supplying condensation nuclei to clouds, which leads to precipitation. In the first experiments precipitation was induced by seeding clouds with tiny dry ice crystals. Dry ice has a temperature of –78° C and water droplets are attracted to the tiny crystals, causing water drops to grow, resulting in precipitation. Silver iodide crystals are also good condensation nuclei and are a cheaper option used today. In South Africa the Bethlehem cloud seeding experiment has been running for many years.

Exploiting weather energy

Wind

Wind has been used for millennia to propel sailing ships and for centuries to drive windmills. The wind pump was invented about one and a half centuries ago. It made large areas with little surface water but reserves of underground water permanently habitable. Tens of thousands of wind pumps are scattered throughout the Karoo and to a lesser extent in other regions of South Africa.

Today meteorological forecasts direct cargo ships away from head winds and stormy seas, and towards favourable winds and calm seas, reducing voyages by up to 10 per cent even though the distance covered is increased. Passenger jet aircraft similarly exploit the wind.

A wind pump

Electrical power generation

Thunderstorms are unpredictable and tapping their electrical power does not seem feasible for this and other reasons (see 'Thunderstorms and lightning'). The power of wind, however, is seen increasingly as a renewable source of energy for the future. Wind generators already make significant contributions to electricity networks in certain European countries and the USA and are being considered for South Africa.

Wind generators do away with the major problems of present power generation technologies – such as nuclear waste management, the pollution caused through burning of fossil fuels, and the long-term silting up of dams used for hydroelectric power generation. The drawbacks of wind generators are only local, such as spoiling natural sites and being hazardous to birds.

The climate of South Africa

Lying within the subtropical belt of high pressure (see 'Winds' below), South Africa has plenty of sunshine and settled weather. It has distinct summer-, winter- and all-year rainfall regions, and four distinct seasons with warm summers and cool winters. Oceans on three sides have a moderating influence on its climate.

South Africa has only half the world's average rainfall. About 65 per cent of the country receives less than 500 mm of rain annually, which is regarded as the minimum for dry-land farming. A high rate of evaporation results in only 8 per cent of its total rainfall being carried off to the sea by rivers, while the world mean is 31 per cent. There has to be a downside to its plentiful sunshine: major climate hazards are extended droughts and sporadic severe floods.

On the extensive central plateau, the altitude keeps average summer temperatures below 30° C, while winter temperatures often drop below freezing point. Compared to other regions of similar latitude, South Africa is colder in many parts. Only the low-lying northern region has an almost tropical climate.

Above: The Atlantic Ocean at Cape Point, in the afternoon

The country has a surface of just over 1,2 million square kilometres. The greater part consists of meadows and pastures and the climate is suitable for cattle, sheep and wheat farming. Arable land makes up 10 per cent of the total surface, permanent pastures 67 per cent and forests and woodland 7 per cent.

Climate temperatures are determined by latitude, altitude and ocean currents. In addition, a climate is dictated by global wind patterns, amount and variability of precipitation, and rates of evaporation. In South Africa these features combine to create four main climate regions: the hot and humid eastern coast, the dry western coast, the Karoo, which has occasional light rainfall, and the north-eastern central territory, which receives extensive rainfall.

Latitude

South Africa extends from 22° to 35° S and almost all of the country is thus extra-tropical. This latitude range results in summer days being about 2,5 hours longer than winter days up in the extreme north, while there is a four-hour difference in the south. In the north (in Pretoria, for example) maximum summer temperatures give way to maximum winter temperatures about 8° C lower.

Topography

The interior is an extensive high plain with an altitude of between 1 000 and 2 000 metres. A narrow eastern coastal plain extends from Cape Town north-east for most of the eastern shoreline. A mountainous escarpment divides the central plateau from the eastern coastal plains, with the KwaZulu-Natal Drakensberg reaching up to over 3 000 metres. In the Drakensberg annual rainfall varies from 1 900 mm on top to 1 000 mm on the plains below. This great escarpment forms a barrier to moist air moving inland from the east.

The Tugela Waterfall in the Drakensberg, dropping from about 3 000 m to 1 400 m in the valley below

Oceans

South Africa has a coastline almost 3 000 km long. The warm Agulhas Current in the east flows southwards from tropical latitudes, while the cold Benguela Current in the west flows northwards from Antarctica. The average annual temperature of Durban Point on the east coast is 18,6° C while that of Port Nolloth on the west coat, at the same latitude, is 14,2° C.

More evaporation occurs from warmer oceans, and air moving inland carries more moisture. Precipitation occurs, and the further the air moves, the less moisture remains. The western region is therefore deprived of moisture moving overland from the east, and moisture moving inland from the cold ocean to the west. The result is a subtropical eastern belt that merges into a semi-desert western area across the width of the country.

The west coast is a semi-desert region with a small temperature range and frequent fog and low cloud. Gale-force winds are frequent, especially on the south-western and southern coastal areas. But the impact of the cold Benguela Current is not the most important factor in determining the climate of the western parts of South Africa. High-pressure systems are created by slowly descending air over a wide area. Descending air inhibits cloud formation and precipitation. More air descends eastward of such systems. The semi-permanent south-Atlantic anticyclone to the south-west of the country therefore has a major impact on the west coast climate. Such systems create deserts on the west coasts of continents worldwide, at latitudes near 30°.

Fog on the west coast

El Niño is a phenomenon that results from interaction between winds in the atmosphere and the surface of the vast Pacific Ocean, during times of shifting patterns in sea surface temperatures. It happens on average every three to five years. When pronounced and persistent, El Niño may have an extreme impact on tropical regions; in middle latitudes its effects may not be as accurately forecasted.

Moisture is carried from the oceans onto the land by winds described below.

Winds

Global conditions

The mean circulation of the atmosphere over the subcontinent influences its climate, while synoptic and smaller-scale disturbances influence its weather. The simplified climatic aspects mentioned here only scrape the surface of a rather complicated and involved topic. The system of cells and winds described below moves somewhat north and south with the seasons and is mirrored in the northern hemisphere

The westerlies

The dominant persistent global atmospheric motion is winds with a westerly component known as the westerlies, which are centred over the mid-latitudes of each hemisphere. Near the earth's surface they occur between approximately 30° to 60° latitude, while in the upper levels they extend further in both directions. In the southern hemisphere westerlies blow from the north-west and are more constant than those of the northern

- Tropical cyclones. These originate in the tropical Indian Ocean and are extremely powerful weather systems, seen as a mechanism to dissipate excess heat energy building up in the atmosphere (see 'Wind storms'). They deteriorate on reaching land and rarely reach South Africa, but cause flooding when they do.

Anticyclones are frequent in warm areas around 30° latitude. They are high-pressure systems created by slowly descending air, extending over thousands of kilometres. Descending air works against cloud formation; the result is fine weather. All droughts, whether short-lived or lasting for years, may be linked to a predominance of anticyclonic conditions. Clouds forming under such conditions are stratiform.

A ridge or wedge is an outward wedge in a high-pressure system towards a low-pressure system. They are usually orientated east to west. A trough is an outward bulge in a low-pressure system towards a high-pressure system. They are usually orientated north to south.

Cold fronts

Cold fronts, invasions of cold air from the south and south-west, and warm fronts occur within middle-latitude or frontal cyclones (see 'Cyclones' above). These are large travelling cyclonic storms up to 2 000 km in diameter. Fronts are boundaries between air masses of markedly different temperature and humidity over a relatively short distance of 50–100 km.

Cirrus cloud ahead of a cold front in the Cederberg, with a cedar tree in the foreground

Cold fronts have a major influence on the weather and climate of South Africa. They occur frequently in winter and typically last for three to five days, travelling up to 10 000 km and sometimes right across South Africa into the northern neighbouring countries. Temperatures at the surface drop markedly, by up to 10° C and more, and it may snow at higher elevations. Cold fronts are associated with distinctive bands of cloud extending far inland. Less frequent summer cold fronts are recognised as the source of severe weather over the interior.

Cold fronts should be understood in combination with warm fronts as a facet of cyclogenesis.

Cyclogenesis

The creation and development of cyclones in the middle latitudes is known as cyclogenesis. Cold and warm air interact to form a rotating weather system. These systems develop in middle latitudes where warm northern tropical and cold southern polar air meet and interact in a complex way under conditions of energy imbalance. Energy radiated by the sun accumulates in equatorial regions, which receive more direct sunlight, and cyclogenesis is the important mechanism in energy exchange between equatorial and polar regions, restoring the energy balance.

The westerlies blow between the warm northern and cold southern air. A polar front (see 'Wind cells' above), a zone about 50 km wide, separates the bodies of cold and warm air. When the temperature gradient increases, the westerlies increase in strength and the flow becomes unstable. This means that when some obstruction creates a small wave disturbance in the airflow, it continues to grow.

Unstable airflow causes a low-pressure system or cyclone to come into being and to continue to grow. Air circulates clockwise around a cyclone in the southern hemisphere. Cold air moving northwards turns clockwise around and into the cyclone.

The cold front encounters warm air moving south behind the warm front. Along the cold front, cold dense air wedges underneath the lighter warm air ahead. Behind the surface position of the cold front, powerful and rapid uplift of this warm air results in the development of cumulus clouds followed by cumulonimbus clouds and thunderstorms.

The warm front encounters cold air moving around the cyclone and southwards, and moves over it, resulting in the uplift of the warm air following the warm front. The uplift is more gentle than with the cold front and therefore stratiform clouds develop behind the surface position of the warm front. Cirrus clouds appear, followed by middle-level clouds and then thick stratus clouds producing widespread rain.

The cold front catches up with the warm front in about 24 hours, which cuts off the supply of warm air to the system, forming an occluded front. The cyclonic storm collapses quickly, leaving inward-spiralling cloud bands. Cold air now moves directly northwards and warm air directly southwards, which tends to wipe out the north-south heat imbalance. The north-south temperature gradient that triggered the cyclogenesis is thereby weakened, but with this the energy exchange weakens and the whole process repeats.

Cold fronts are more sharply defined and also more noticeable by humans because the land is warm anyway. Warm fronts occur mostly over the sea and do not affect the country.

Rain winds

Moist air is carried from the oceans to the land by moisture-laden winds. Various mechanisms cause air to rise over land, its moisture to condense into clouds and precipitation to follow (see 'Thunderstorms and lightning').

The South Indian Ocean anticyclone to the south-east of the country, the South Atlantic anticyclone to the south-west of the country, and the continental high over the country, broken in summer by convection, are semi-permanent elements of the subtropical high pressure belt that dominate South African weather.

In summer, moisture is carried from the oceans to the land by winds originating in these anticyclonic regions:

- Winds circulating around the northern part of the South Indian anticyclone travel a long way over the ocean, picking up a lot of moisture before reaching the country from the south-east. Over land, lift is created, followed by rain.
- Usually winds from the South Atlantic anticyclone follow a much shorter path over the ocean before reaching land and are therefore less moist. Sometimes the anticyclone moves south, resulting in a longer path over the ocean and more moist winds. This may be followed by rain over the south-eastern parts of the country.
- A convergence zone results where these winds meet the South Indian Ocean anticyclone winds from the east. The western winds wedge underneath the more moist eastern winds, creating lift and rain.
- Winds from the South Atlantic anticyclone follow a longer path across the ocean and reach Angola from the south-west. A convergence zone results where these winds meet the South Indian Ocean anticyclone winds from the east. This has a major influence on the weather of Namibia and northern Botswana.
- The warm Kalahari deflects moist eastern winds southwards, towards the southern Free State.
- Coastal lows lead to fog on the west coast and cloudy conditions and drizzle on the east coast, in both cases to the west of the low.

The rain-producing disturbances of tropical origin peak in summer, while those of westerly origin peak in autumn and spring.

Turbulent ocean

Climate regions

Borders between climate regions, although sharply defined on a map, should be seen as broad transitional zones. Published maps deviate markedly due to fine-tuning of the defining parameters. A climate region subdivision of the country may range from a few broadly defined regions highlighting major climatic parameters, to more finely defined regions highlighting minor climatic differences. Plants adapt to climate and a vegetation map is often similar to a climate map of the same area.

In the following sections, covering eleven climate regions, adjoining regions have been grouped together to highlight trends.

Mountain ranges constitute the main climate divides and have been used by the SAWS in partitioning South Africa into climatic regions. Secondary considerations are rivers and political boundaries, ans also the basis used for partitioning the interior plateau.

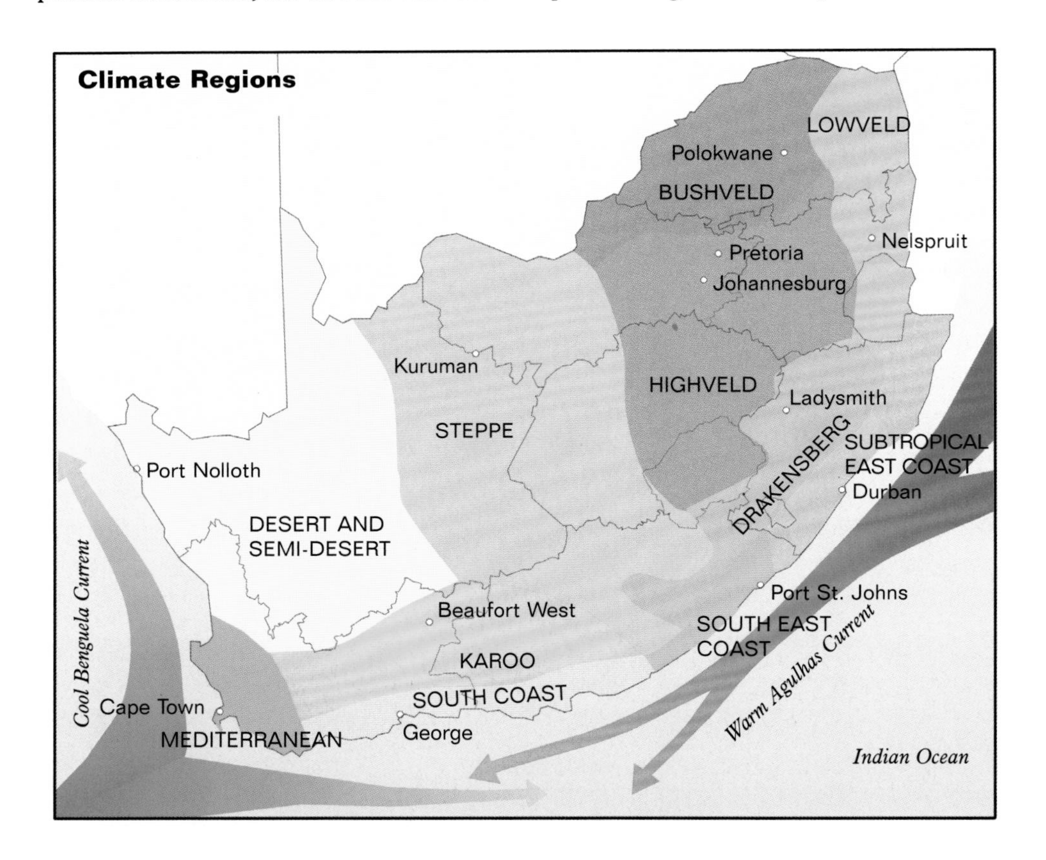

Rainfall patterns

The major climate regions have distinct rainfall patterns:

- Summer rainfall from December to February in the northern regions, e.g. Pretoria, Bloemfontein, Durban, Zeerust, Musina.
- Late summer rainfall from mid-January to mid-April in the central east-west belt, e.g. Kimberley, Upington, Cape St Lucia.
- Rainfall throughout the year in the south-east, e.g. Swellendam, Uniondale, East London.
- Winter rainfall from June to August in the south-west, e.g. Cape Town, Port Nolloth, Montagu.

Average rainfall, however, may differ vastly within a region, influenced by mountain ranges.

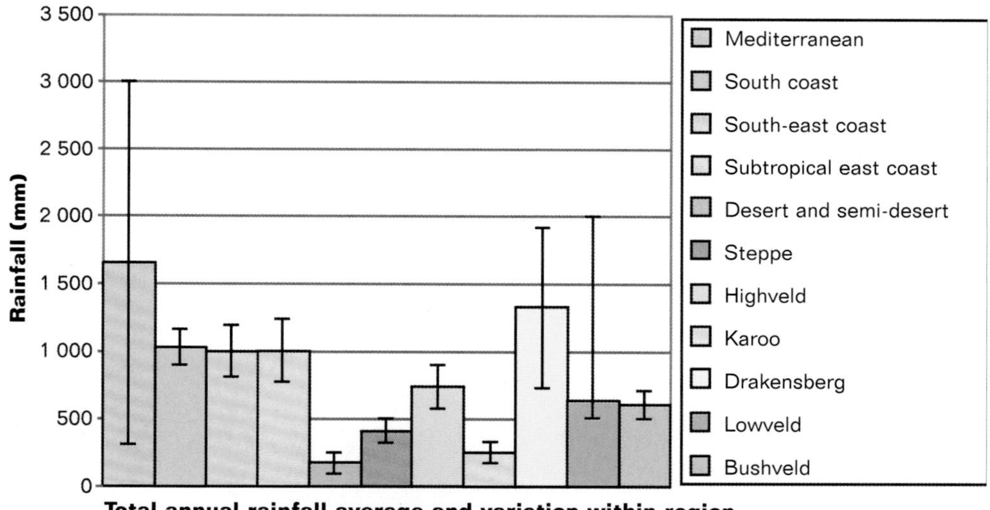

Total annual rainfall average and variation within region

Up the eastern coast

Moving north along the eastern coast from Cape Town towards Durban, rainfall changes from winter to all-year to summer.

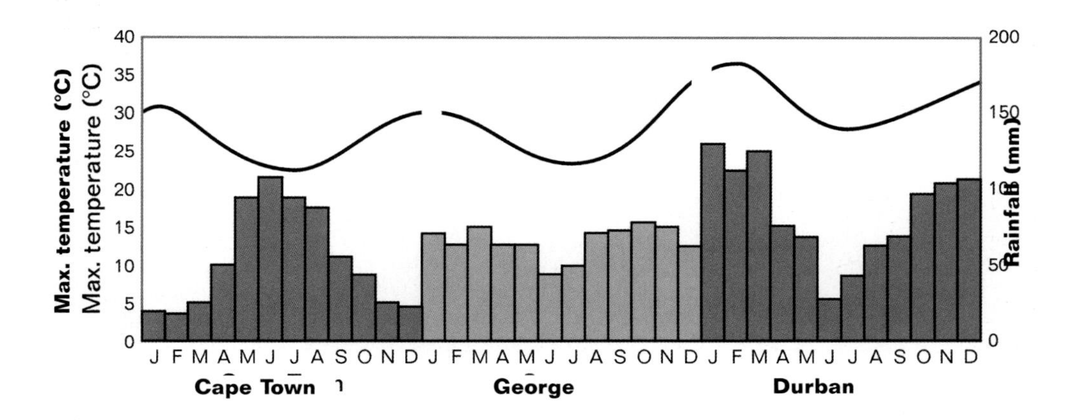

Mediterranean

The south-western corner of South Africa below the escarpment has a Mediterranean climate. It experiences warm to hot summers and mild to cool winters, while most rain and the occasional snow on the mountains fall in winter.

Proximity to the ocean dominates the climate. Furthermore, the region is so far south that it is influenced by the Southern Ocean circulation pattern of moist air. The high pressure systems move somewhat north during winter, allowing the westerlies to reach the southern tip of the country.

The typically low nutrient content of the soil in a Mediterranean climate zone favours the growth of shrubs like fynbos in the Western Cape. Most plant growth occurs in spring following the winter rain. The rainfall pattern favours wine-farming.

Farm near George (south coast)

South coast

The south coast region is a long, low-lying and narrow stretch ranging from the Western Cape up to the southern border of KwaZulu-Natal, and bordering on the Indian Ocean. It is a wet region, with rainfall occurring throughout the year but peaking during spring and autumn. The climate of the region is determined by its low altitude (temperatures decrease with altitude) and the warm Agulhas Current. It is a temperate climate, especially near the coast, with humid and warm summers, and mild winters.

South-east coast

The south-east coast region extends northwards from the south coast region. It has a temperate climate with humid, warm summers and mild winters. Warm Berg winds are experienced in winter. Rainfall is spread throughout most of the year, with only 65 per cent of the annual rainfall falling in summer. Rainfall decreases inland. The region is sometimes combined with the south coast region above for a simplified regional climate subdivision.

Subtropical east coast

The subtropical east coast region extends northwards from the south-east coast region. It experiences hot, humid summers and dry, frost-free winters, with 70 per cent of its rain falling in summer. A subtropical climate zone is characterised by lower temperatures and more seasonal rainfall than the equatorial tropics. Wet and dry seasons are of similar length.

Nsumo Pan, Mkuze Game Reserve (subtropical east coast)

West to east across the interior

Moving across the country from west coast to eastern highveld, arid regions gradually give way to summer rainfall regions.

Desert and semi-desert

This arid climate region is characterised by desert conditions where evaporation exceeds rainfall. Average rainfall is less than 250 mm per year, with no rainfall during some years. Large daily and seasonal temperature fluctuations occur. The climate of this region is determined by its low altitude, the cold Benguela Current and the semi-permanent South Atlantic anticyclone to the south-west of the country. Plant and animal life are highly adapted – for example, the flowers of Namaqualand. The Benguela Current is responsible for the Cape upwelling system, which produces nutrient-rich waters favouring plankton growth at the bottom end of the food chain, leading to an abundance of fish on which the country's fishing industry is based.

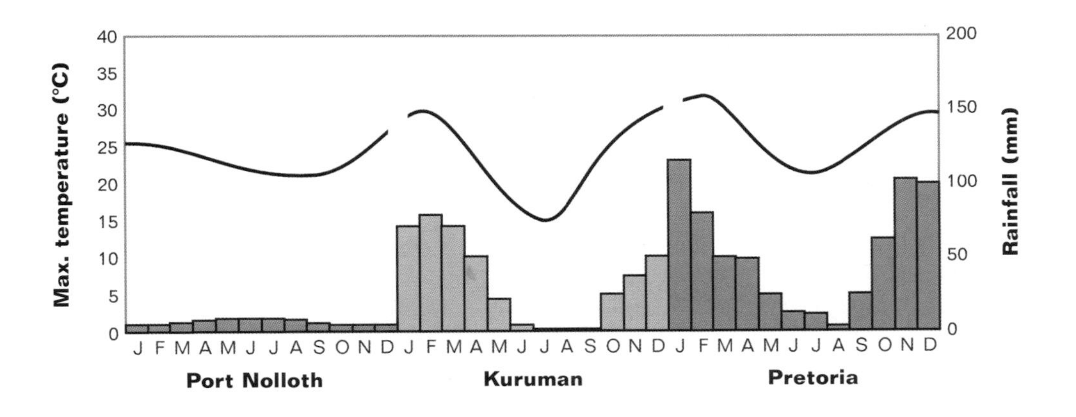

Kgalagadi Transfrontier Park (semi-desert)

Steppe

The steppe covers a broad band from the Limpopo River in the north to the Karoo in the south. It borders on six of the other regions. It occupies the fairly dry interior, separating the highveld from the Karoo and arid north-west. A semi-arid climate zone is characterised by low rainfall, too little to support forests but enough to prevent desertification. Dominated by large expanses of grassland known as veld in South Africa, it covers a large tract of the inland plateau. This region is sometimes combined with the Karoo to its south and the semi-desert region to its west in a simplified regional climate subdivision.

The veld near Kimberley (steppe)

Highveld

Gauteng and Free State (and Lesotho) fall within the highveld region, which is characterised by warm summers and mild winters with frost. Most rain is produced by summer thunderstorms. Occasional cold fronts are experienced, moving in all the way from the south-western seas. Snow falls on higher areas only rarely during extreme cold fronts.

Near Ventersdorp (highveld)

Rain shadow and rain belt

When moist air moves inland from the oceans it is lifted by topography. Clouds form, sometimes followed by precipitation, on the mountain slopes facing the ocean.

Mountains impact especially on two climate regions of South Africa. The Cape mountains deprive the Karoo of rain, by restricting rain to the ocean side of the mountains. The Drakensberg region benefits from rain on the ocean side of the Drakensberg mountains.

Karoo

The Karoo region lies inland from the wet Mediterranean and south coast regions, from which it is separated by mountain ranges creating a rain shadow. At a higher altitude on the inland plateau, it is a semi-arid region with low and unreliable rainfall. Periodic long droughts occur, as do occasional floods. Midsummer high temperatures average above 30° C and often push 40° C, while winter nights are often freezing. The invention of the windpump one and a half centuries ago made this region permanently habitable for humans.

The Karoo

The Drakensberg

The Drakensberg climate region occupies the KwaZulu-Natal Drakensberg mountains, its foothills and lower plains extending up to the east coast region. The higher areas have a mountain climate, which is characterised by much lower temperatures than low-level regions of similar latitude, due to its altitude. Regular snowfalls occur in winter and sometimes even in summer on the high mountains. Summer thunderstorms bring most of the region's rain. The extreme altitude range results in subregions with markedly different features: for example, average annual rainfall varies from 1 000 mm in the lower regions to 1 900 mm in the mountains, making a one-figure average meaningless.

In the lower regions, the rainfall and temperature fluctuations are similar to those of the bordering subtropical east coast, but temperatures are a few degrees colder and rainfall is caused mainly by thunderstorms.

Below the Drakensberg

The lowveld and bushveld

Lowveld

The lowveld occupies the area below the Mpumalanga escarpment, bordered by the Limpopo River in the north, Mozambique in the east and Swaziland (which falls within the region) in the south. Most rain occurs during the hot and humid summer. Moist air originates in the tropics and Indian Ocean. Rainfall increases towards the escarpment. Maximum temperatures often rise towards 40° C, high above the average in the mid-twenties.

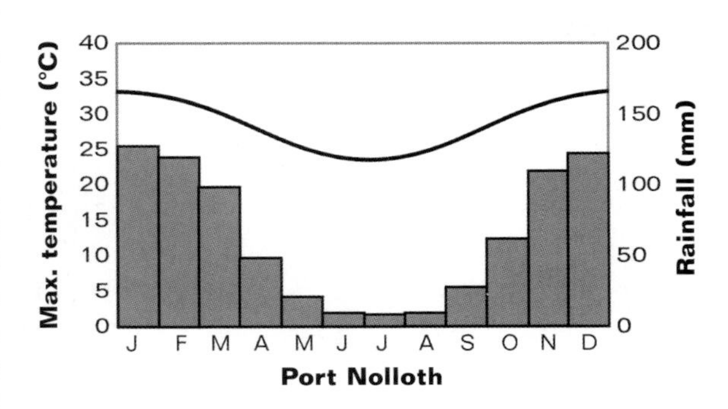

Near Letaba in the Kruger National Park (lowveld)

Boabab tree north of the western Soutpansberg (bushveld)

Bushveld

The bushveld occupies the north-western part of South Africa bordering on Botswana and Zimbabwe, with the Limpopo River as boundary. Hot summers with maximum temperatures moving into the lower thirties are typical. Winter nights are cold to sometimes very cold. Heavy thunderstorms are common and produce most of the annual rainfall. The region does not differ markedly from the lowveld region, except in rainfall, and they are sometimes combined in a simplified regional climate subdivision.

Climate change

Research into past climates indicates that over time the climate of the world is highly erratic but that this is superimposed on periodic climate changes, with periods of thousands of years linked to orbital parameters of the earth. Today humankind influences the climate through the release of gases into the atmosphere. The extent of this influence is hotly debated and more research is needed to separate natural climate variability from trends due to human activities.

The Gaia hypothesis

The Gaia hypothesis is the idea that all life forms on the planet are part of a single living organism, the earth. Like the cells and organs in the human body, the life forms of earth have co-evolved to produce and sustain optimal conditions for the growth and prosperity of the larger whole. The implication is that significant human meddling in the state of the earth will invoke a drastic natural corrective response.

Global warming

Without the greenhouse effect (see also 'The radiation balance') containing some heat the earth would be much colder, too cold to support life in general. However, an artificially increased greenhouse effect causing temperatures to rise dangerously is of great concern.

Increasing concentrations of greenhouse gases are likely to cause global temperatures to rise, precipitation to increase in high latitude areas and decrease in tropical areas, and sea levels to rise due to the melting of glaciers as well as the expansion of warmer ocean waters. Such observations are indeed made. In the last century the average land surface temperature rose by about 0,5° C, precipitation increased by about one per cent over the world's continents and declined in many tropical areas, and the sea level rose approximately 15–20 cm worldwide. In Africa the icecaps of Mount Kilimanjaro and Mount Kenya have visibly retreated during the timespan of a human lifetime.

The Milankovich theory

According to the Milankovich theory, the shape of the earth's orbit around the sun and the tilt of its rotational axis relative to the orbital plane undergo cyclic changes over thousands of years, changing the distribution of solar radiation falling on earth, influencing its climate dramatically.

The earth is in an elliptical orbit around the sun, with the sun centred in one of the two elliptic poles. In a six-month cycle it reaches the opposite points on the ellipse nearest to and farthest from the sun. The nearest point is reached around 2–3 January during the Southern Hemisphere summer, and the farthest point around 5–6 July during the northern hemisphere summer, resulting in 7 per cent less summer sunshine there.

The earth's orbit varies from near circular to highly elliptical every 100 000 years and a further cycle of 433 000 years is superposed on this. As the orbit becomes more elliptical the northern hemisphere receives less and the southern hemisphere more radiation throughout the year, although the annual total remains the same. In the extreme elliptical state the intensity of sunshine may vary by up to 30 per cent. The 100 000-year major ice age cycles correlate with this variation. We are presently within an interglacial period, lasting 10 000 years, between major ice ages.

The tilt of the earth, presently 23,5°, varies from 22 to 24,5° over a 41 000-year cycle. The changing tilt is known as nutation. A stronger tilt increases and a slighter tilt reduces the difference between summer and winter. For the past 10 000 years the degree of tilt has been declining, slowly causing cooler summers and warmer winters.

The earth wobbles like a gyroscope, its rotational axis tracing a circle every 23 000 years. This causes the seasons to drift and the point nearest the sun, presently reached in early January, will in about 11 000 years be reached in July. This applies to years measured in terms of solar orbits and not seasons. This movement is therefore also known as precession of the equinoxes. The effect will be warming of both southern hemisphere winters and northern hemisphere summers.

Changes in the tilt of the earth as well as precession of the equinoxes are deduced from the altitude and circular movement of the celestial poles in the starry night sky. These movements were noticed and measured by ancient Greek astronomers.

The variations in orbital movements are not enough to effect drastic climate changes on their own, but are amplified by other phenomena. When the climate turns colder, ice fields on the surface of the earth increase in size, reflecting more sunlight. The opposite happens when the climate turns warmer, as at present. Glaciers retreat and large chunks of Antarctic ice banks break up. Recently a chunk with a size of 3 250 km^2 broke away from a 220-metre-thick, 12 000-year-old ice bank, forming thousands of icebergs which will eventually melt.

Thunderstorms and lightning

Convection

Convection – the warming of air close to the earth, causing it to rise – is a major force in thundercloud development. Since oceans absorb less heat and remain cooler, by far the greatest number of thunderstorms develop over land.

At any moment roughly 2 000 storms are in progress worldwide. With up to three cloud-to-ground flashes per minute produced by the typical storm, that makes a total of about 100 every second. The incidence increases towards the tropics. Kampala in Uganda has the highest recorded incidence of 242 thunderstorm days per year (calendar days during which thunder is heard once or more at a specific location).

Above: Isolated thundercloud in the afternoon, Pretoria. The image shows a fully developed thundercloud with typical anvil top spreading out sideways on reaching the stratosphere

In central Africa thunder is heard on about 150 days and in most of Europe on about 11 days per year. In South Africa thunderstorms are common and typical in the north-eastern summer rainfall regions but rare elsewhere. In and around Johannesburg thunder is heard on 50–60 days per year and in the Cape Peninsula only about 12 times per year.

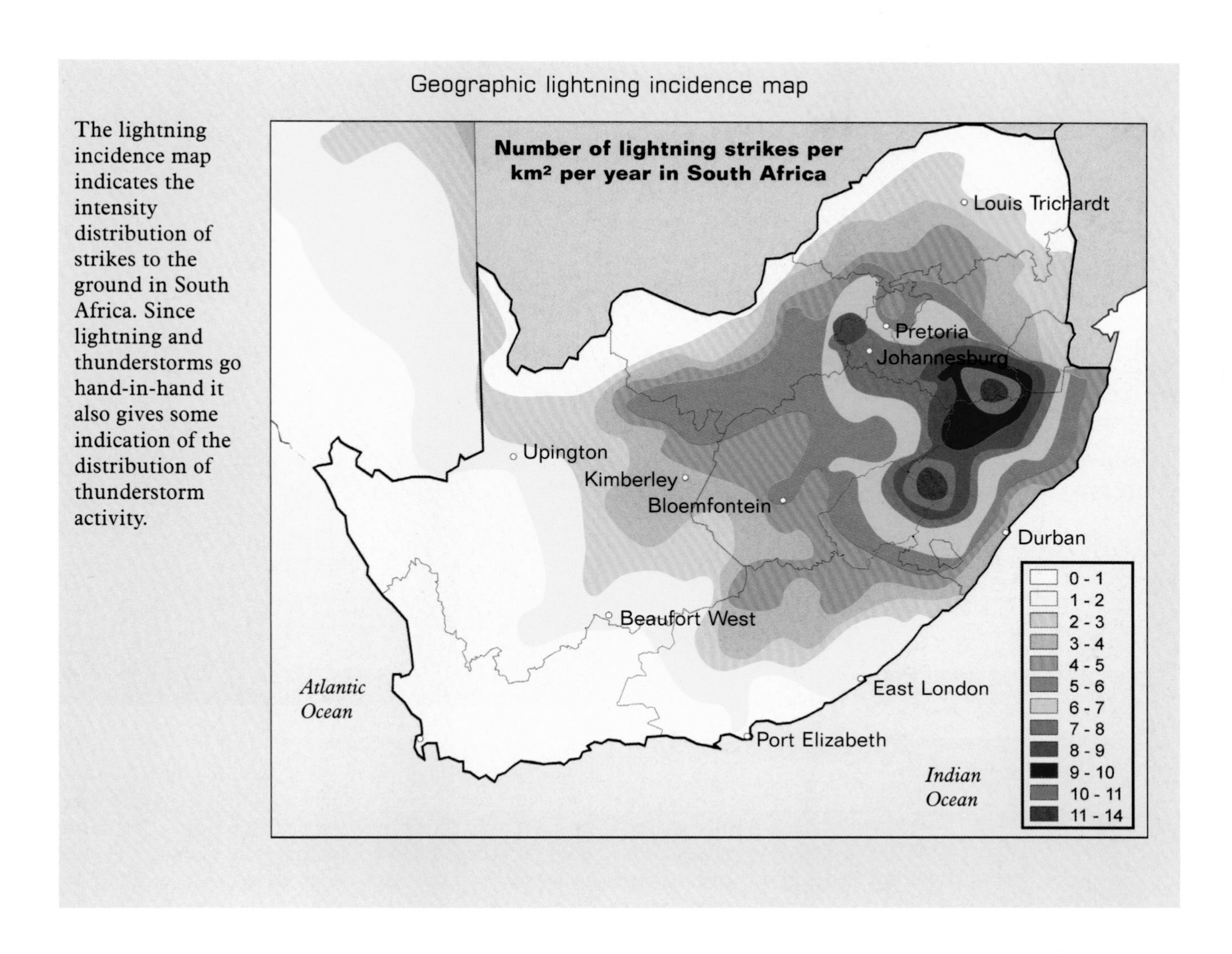

The lightning incidence map indicates the intensity distribution of strikes to the ground in South Africa. Since lightning and thunderstorms go hand-in-hand it also gives some indication of the distribution of thunderstorm activity.

The life cycle of a thunderstorm

The main ingredients of thunderstorm formation are moist air, lift and instability. Moist air results from the evaporation of water. The sea contributes most moisture to the earth's atmosphere. Lift is caused by:

- *Convection*. Warm air is lighter than cold air and its buoyancy makes it rise.
- *A moving body of cold air*. A body of cold air bumping into a body of warmer air wedges underneath, forcing it up. Cold fronts are large-scale and thunderstorm downdraughts small-scale examples.
- *Wind over mountains*. Air is forced up by the topography.
- *Convergence*. Air moving into a region from opposing directions cannot accumulate and is forced upwards. Convergence is usually accompanied by divergence higher up, which distributes the air carried up.

Air is said to be unstable when it does not return to its previous condition after being disturbed. Instability occurs when a layer of warm and moist air is situated below a layer of cooler, drier air, or when air close to the ground is heated by a warmer earth. It is fuelled by the release of latent heat by moist air when condensation occurs on reaching middle levels, and again when water droplets freeze at higher levels.

Rain, hail, strong winds, microbursts and sometimes tornadoes, accompanied by lightning and thunder, are unleashed by a thunderstorm.

A regular patchwork pattern of cumulus cloud seen over the Mooi River

Convective storms

On a hot day the sun heats the earth. The layer of air near the ground is warmed and it starts rising. The air rises in discrete vertical channels about a kilometre in diameter known as Bénard cells, spaced in honeycomb fashion over the heated terrain. The warm air is displaced by colder air descending in the spaces between the cells. Cumulus clouds form in a patchwork pattern. Convective storms usually develop after midday, to disappear in the late afternoon. Being isolated, with few active cells they are ideal for observing lightning.

Frontal storms

When a body of cold air moves into a region of warm air it wedges in underneath the warm air, forcing it upwards. The advancing head of a large body of cold air is known as a cold front. The frontal line is known as a squall line and the associated storms as squall-line storms. The cold front adds moisture, lift and instability. These storms are more violent than convective storms, producing stronger winds and bigger hailstones.

Supercell storms

Supercells are storms that form at the end of squall lines. One cell may be tens of kilometres across and dominate a storm. They can last several hours because the cold front continuously supplies cold air at middle levels, enhancing instability. They are even more violent than frontal storms.

Cells

A thundercloud is characterised by regions of strong upward and downward air currents known as cells, which develop through three successive stages. An extensive or multi-cell thundercloud has several cells in various stages of development.

An isolated Cumulonimbus cloud has a diameter of about 8 km while the diameter of the cell inside is about 1,5 km. The cloud head reaches up to a height of 10–16 km above the ground, much higher than Mount Everest (just below 9 km above sea level).

Single- or multi-cell storms form as a result of convection alone. Multi-cell storms are clusters of single-cell storms feeding off each other's downdraughts, which cause surrounding warmer air to rise, much like the effect of a cold front.

In the developing or Cumulus stage an updraught exists throughout the cell. The mature stage commences when the first rain reaches the ground, downdraughts develop and the cell is electrically active. The electrically active state lasts some 20 minutes. Strong winds, heavy rain and sometimes hail occur at ground level. In the dissipating stage the downdraught spreads horizontally throughout the cell and diminishes, and a steady shower ensues. All three stages last about an hour. The Cumulus stage lasts 10–15 minutes, the mature stage 15–30 minutes and the dissipating stage about 30 minutes.

Developing stage of a cell

- In the troposphere, the lower layer of the atmosphere, the temperature decreases with increasing height at a rate of 6,5° C per 1 000 m (wet adiabatic lapse rate).
- As air is warmed by the sun it becomes less dense and rises in a process called convection.
- The rising air cools with decreasing pressure at increasing height. Moisture in the form of water vapour in the air condenses when dewpoint is reached to form clouds.
- The Cumulus congestus stage forms, and the cloud looks like a white mountain.
- Instability is necessary at the middle and upper levels of the troposphere, if there is to be further development. This is provided by the release of latent heat when moisture in the air condenses and water drops freeze.
- When the top of the cloud reaches the tropopause, between the troposphere and the stratosphere, where the temperature starts to rise with increasing height, it reaches a level where it is no longer warmer and lighter than the surrounding air. The cloud top stops rising and spreads out, pushed out by ever more air still rising from below. The temperature at the top is well below 0° C and the water exists as ice crystals.
- The Cumulonimbus cloud has formed, with the cloud top looking like an anvil and the cloud base dark, and it starts to rain.

Mature stage of a cell

- The mature stage starts when the first rain reaches the ground.
- A downdraft develops as cooled air, more dense than the surrounding air, falls back to earth, also aided by falling rain.
- The upward and downward air movements help to create opposite electrical charges collecting at the bottom and top of the cloud. Lightning follows.
- Electrical discharges also enhance coalescence (drop formation) within the cloud.

Dissipating stage of a cell

- Downdrafts eventually overwhelm the supply of warm air, and the storm dissipates.
- Cirrus and small Cumulus clouds remain.

Accompanying winds

The strong updraughts in the thundercloud suck in air, causing surface winds to blow towards the storm from some kilometres away. Near a thunderstorm the cold downdraughts blow away in strong gusts of wind at the surface. Thunderclouds move horizontally at speeds generally in the order of 30 kph with the prevailing upper winds. Cold air extends far to the rear of the storm and may subside for up to an hour or longer at ground level. The drop in temperature from uncomfortably hot to pleasantly cool is typical in South Africa as in other subtropical regions.

The dynamics of a lightning flash

The dynamic stages of the spectacular cloud-to-ground flash are described below, in the common case where the cloud base carries a negative charge. 'Lightning flash' is used to describe the complete process from the initiating stepped leader (see below) through multiple strokes, together typically lasting up to half a second and often longer in South Africa. Much happens in less than one second, so lightning flashes are described in terms of thousandths of a second or milliseconds (ms) and even millionths of a second or microseconds.

The build-up of electrical charges

The build-up of huge opposite electrical charges in a thundercloud (Cumulonimbus) remains somewhat of a mystery. Violent updrafts and downdrafts in the thundercloud during a storm play an important role in creating charged particles by means of collisions and friction, and in transporting them in opposite directions. Positively charged ice particles are deposited in the top region while negatively charged water droplets sink to the bottom of the cloud. A strong charge opposite to that in the cloud bottom also builds up in the ground below and follows the moving cloud like a shadow. Lightning strikes into this region, which is several times wider than the cloud's horizontal diameter. The opposite charges strongly attract each other but the flow of electricity is inhibited by the air, a very poor conductor, in between.

In stratiform clouds updraught velocities are small and rain without lightning occurs.

Conducting region

A greater part of the lightning stroke exists inside the cloud. A conducting region is formed inside the negatively charged region of the thundercloud. Small sparks forming streamers merge into a branched structure converging downwards. Negative charge is drained from here by the stepped leader prior to and during its descent.

The stepped leader

Once the charge build-up in the cloud becomes sufficient to overcome the insulating layer of air below, negative charge begins to flow from the lower regions of the cloud, hesitantly and tortuously zigzagging downwards, tentatively branching into many offshoots before continuing. This stepped leader propagates slowly compared to the speed of light. It is not extremely bright and therefore unlikely to be seen, the false branches only becoming prominent when lit up by the return stroke, when the main branch becomes the prominent lightning channel.

The diameter of the stepped leader is between one metre and 10 metres. Individual steps are 10–80 metres long and intervals between steps range from 50 to 90 microseconds. The stepped leader advances at a fairly constant overall velocity that suggests the presence of a pilot streamer advancing continuously ahead.

Positive streamers

When the stepped leader reaches about 50–20 metres above the ground, positively charged streamers from tall grounded objects such as trees reach up to meet it. Once a connection is made a highly conductive ionised channel is established between the cloud and the ground and the return stroke follows. The leader channel consists of a conducting core surrounded by a corona envelope, a glow discharge to the surrounding air

The return stroke

A luminous flow propagates from the ground up to and into the cloud at up to half the speed of light, too fast for the eye to follow. This is the most visible stage of lightning, and is associated with the most audible stage – loud thunder.

The return stroke is a heavy flow of electrical current that neutralises the opposite charges between cloud and ground in a very short time. It brightly illuminates the ionised channel as well as the false branches created by the stepped leader and establishes a highly ionised channel with a diameter of around 16 cm that remains conducting for some time.

In this image the bright luminous channel, the return stroke, is surrounded by a wide luminous electrical corona. The image was captured from Pretoria, looking north at night. The Union Buildings may be seen through the rain at the horizon

The ionisation results from the extreme heat generated, causing atoms to lose one or more electrons. The core of this channel has a diameter of only a few millimetres, based on the size of burn marks left on lightning rods.

Although the region of high current and luminosity moves upwards, electrons at all points along the channel move downwards.

The sound of thunder

The current flowing during the return stroke superheats the air to 22 000° C, more than four times hotter than the surface of the sun, so fast that it has no time to expand. This produces a shockwave, a major disturbance in the air travelling faster than the speed of sound. It compresses the surrounding air and soon decays into an acoustic wave, creating the loud thunderclap heard up to 30 km away. Thunder generally has a frequency of 50 cycles per second. A much lesser sound is generated by the stepped leader and by upward streamers emanating from tall objects on the ground. Sound is also generated during the sudden build-up of heavy current in the return stroke at the base of the flash. The sound heard during a lightning stroke is made up of these different sounds and what is heard depends on the distance between the listener and the point of impact.

When lightning strikes at a distance, sound waves from various lengthy segments of the channel undergo acoustic interference on their way to the listener, which is heard as rumble.

When lightning strikes with a sound resembling tearing cloth preceding a sharp thunderclap, it struck nearby, within a couple of hundred metres. The sharp clap results from the sound generated all along the lightning channel reaching the observer almost simultaneously. It has been suggested that the tearing sound arises from a multitude of points situated close to the strike point from which streamers are emitted.

Lightning research in South Africa

The Bernard Price Institute of Geophysical Research

Basil Schonland completed his PhD in physics shortly after the First World War at the famous Cavendish Laboratory of Cambridge. There he rubbed shoulders with a dozen or so present and future Nobel Prize winners. The laboratory was at the height of its fame with its director, Ernest Rutherford, having deduced the existence of the neutron and named it nine years before its experimental confirmation. But when Schonland married a South African the couple decided to settle in South Africa. Once back he realised that he would not keep up with science in Europe, being so far away (this was before jet planes and the www) and he looked for a field of research favoured by local circumstances. He was awed by thunderstorms and lightning and soon became a world authority on lightning. This was in his capacity as director of the newly founded Bernard Price Institute of Geophysical Research at the University of the Witwatersrand, the fame of which grew with his stature as scientist. Here he was later joined by Dawie Malan, a former student of his, and together they contributed significantly to the unravelling of the secrets of lightning during the first half of the twentieth century. Their main tool was further developed Boys cameras. This camera utilises a rapidly rotating lens or film that enables the dynamics of the lightning flash to be spread out in time on the film, revealing its various stages.

Schonland and Malan were by no means the only leading lightning experts of their time and other countries have since taken lightning research into new heights, employing new tools, for example lightning sensors aboard satellites.

The CSIR

In later years lightning research moved to the Council for Scientific and Industrial Research in Pretoria, founded by Schonland at the request of Jan Smuts. A notable result of research done here is the lighting incidence map of South Africa featured above.

Other organisations

The SABS (South African Bureau of Standards) is mainly concerned about safeguarding buildings and industry. Eskom's main concern is the protection of power transmission lines.

When lightning strikes within 100 metres a click is heard, then a whip-like crack, followed by continuous rumbling. The click arises from streamers emitted by trees or buildings reaching up to the incoming stepped leader, and is heard first because of being nearer the observer. The sharp crack results from the sudden build-up of heavy current in the return stroke at the base of the flash. The rumble results from regions higher up – and therefore progressively further in the lightning channel – and subsequent strokes.

Distant thunder

The distance to lightning several kilometres away may be deduced from the speed of sound in air, which is much slower than the speed of light. Sound needs three seconds to travel one kilometre in air, so that the time between seeing a flash and hearing the thunder, divided by three, indicates the distance. This is an approximation only, since the speed of sound in air is dependent on the air density, which diminishes with height above sea level, and depends on local high and low pressure conditions and the moisture content of the air.

Thunder may be heard up to 30 km away, equal to counting to 90 from seeing the flash. Closer than about 8 km this formula is less reliable, since the flash might originate nearly overhead and travel some distance horizontally before joining up with an existing downward channel further on. The horizontal part of a flash is often longer than the vertical part. The strike point then was not as close as calculated.

During a multiple stroke flash, successive strokes will be heard and be longer, producing longer rumblings.

Multiple strokes

Multiple strokes cause the flickering display often seen. If sufficient charge exists in a thundercloud many strokes may occur during a single lightning flash, with subsequent strokes making use of the lightning channel established by the first stroke. A lightning flash typically lasts 200–400 ms in South Africa, with 40–50 ms separating the strokes, indicating five to 10 strokes per flash. In severe storms a flash may last for as long as a minute. Up to 31 strokes have been photographed within 600 ms, indicating 20 ms separating individual strokes.

While a stepped leader precedes the initial stroke, dart leaders precede subsequent strokes.

The dart leader

The leader to subsequent strokes advances to earth continuously and not by steps, through the main channel prepared by the first stroke, ignoring branch channels left by the first stroke. It resembles a fiery dart some 50 metres long, moving too fast to be seen.

Following the first return stroke, the branched top of the conducting channel (see 'Conducting region' above), where the stepped leader originated, becomes positively charged. A strong electric field is created between the branched top and the surrounding negatively charged cloud. This electric field enables positive streamers known as J-streamers to advance upwards and outwards, making a further and higher region of the cloud conductive. Negative charge collects into this region. J-streamers cause the cloud to remain faintly luminous between strokes.

Sufficient charge may collect to initiate another stroke. Then the dart leader moves down the existing channel paving the way. The return stroke follows when the dart leader reaches the ground.

The sequence of J-streamers, dart leader and return stroke is repeated, with the dart leader originating ever higher in the cloud, until the negative charge has been exhausted. Multiple strokes occur, successively originating roughly half a kilometre higher in the cloud.

Effect on the ground

The charge is dissipated into the ground, causing a ground current to flow outwards from the point of strike.

Effect on rain

A lightning flash greatly enhances coalescence, the merging of tiny water droplets into larger drops, initiating a downpour which reaches the ground within a minute or two of the flash. Water droplets in a thundercloud carry an electrostatic charge. The J-streamers (see 'The dart leader') associated with multiple strokes leave droplets with an opposite charge in the streamer channels. These are attracted to unaffected neighbouring droplets and they coalesce, forming larger drops whose weight overcomes updraught forces.

Positive cloud-to-ground flash

When the anvil top of a thundercloud, typically carrying positive charge, is blown some distance away from the cloud below, or blown lower, or develops next to a mountain, lightning may strike directly from cloud top to earth some distance from the storm. Carrying positive charge from cloud to ground, this is a positive lightning stroke. It is less common than the usual negative flash from cloud base to earth and more powerful due to its longer path, which is overcome only after a greater build-up of electrical charge. Usually occurring towards the end of a storm, it is popularly known as the 'parting shot' and might also be one manifestation of the 'bolt from the blue' striking unexpectedly (the other being intercloud lightning losing its way to the ground – see 'Intercloud lightning' below). About 10 per cent of cloud-to-ground flashes are positive and 90 per cent negative.

The dawn of the science of electricity

In 1600 William Gilbert, physician to Queen Elizabeth I, initiated research into static electricity. Materials like amber and glass, when rubbed, attract light bodies and he called them electrics after the Greek electron for amber. During the next century and a half slow progress was made. Newton gave some attention to the subject, remarking on the similarity between short electric sparks and the lightning discharge.

In January 1746 Pieter van Muschenbrock of Leyden invented the Leyden jar, the first condenser in which electricity could be stored. At the same time, Benjamin Franklin turned his attention to the link between electricity and lightning, becoming the first to proceed from speculation to experiment and conducting his famous kite experiment. By 1750 he had proposed the lightning rod to protect buildings and ships. Experiments proposed by him and conducted in France in 1752 confirmed that lightning was an electrical discharge. The lightning rod soon became widely implemented. Franklin declined to patent or otherwise profit from his invention.

In 1799 the voltaic cell of Volta was produced. It provided the first source of continuous low-voltage electrical current and set in motion the great electrical developments of the nineteenth century. But the science of thunderstorm electricity stagnated for another century and a half until high-speed photography, radar and other tools became available. This was the Schonland era.

Under rare conditions the charges in the cloud may be reversed, when the stroke from the bottom of the cloud to the ground also becomes a positive flash, delivering positive charge. This is associated with tornado development. A positive flash propagates in the same way as a negative flash. Positive flashes rarely consist of more than one stroke.

Catalogue of lightning phenomena

Although various lightning phenomena are described on the following pages, the fundamental kinds of lightning are cloud-to-ground, intracloud (in-cloud), intercloud (cloud-to-cloud) and cloud-to-air. They are distinguished by different mechanisms of electrical breakdown, through which air becomes highly conductive resulting in a lightning flash, and which is described above for the cloud-to-ground flash. Strange forms of lightning high above thundercloud tops have lately been discovered and investigated.

Cloud-to-ground lightning

Cloud-to-ground lightning occurs between the base of the thundercloud and the ground. It is initiated by a discharge, called the stepped leader, travelling from the base of the thundercloud to the ground. Since the stepped leader progresses in rapid steps lasting less than a millionth of a second, it is not generally seen by the eye, but captured on film, as the image shows. When an advancing branch contacts the ground, the visible lightning flash occurs, shown by the thicker and brighter part of the image. The bright luminous channel, called the return stroke, originates at the ground and then moves up to the cloud. It moves too fast for the eye so that it appears as if the whole channel becomes luminous simultaneously.

The dynamic process of cloud-to-ground lightning is described in some detail earlier.

Cloud-to-ground lightning seen from Erasmusrand, looking south-south-west towards Johannesburg, not long after sunset as the storm intensified and moved in closer

Lightning in a rain shaft

Cloud-to-ground flashes usually occur inside or near a rain shaft. The front of a moving storm is electrically more active. Watching an advancing storm is therefore more dangerous, but better for photography since flashes towards the front of the storm are obscured less by rain.

The image shows a distinct rain shaft with a cloud-to-ground flash inside.

Lightning in a shaft of rain seen from Erasmusrand, looking west in late afternoon

Multiple strokes

Once the channel from cloud to earth has been established by the first bolt of lightning, several more strokes may occur through this channel, for up to half a second or even longer, but seldom lasting more than one second overall. Multiple strokes are produced with the dart leader originating ever higher in the cloud. The strokes are typically 40–50 ms apart. A photographic exposure of 1/30 s (about 30 ms) or less may therefore fall in between and miss recording a stroke.

Multiple strokes are seen by the eye as a flickering, but the strokes are lost in still photography since they overlap and re-expose the same part of the film. A camera may be deliberately swung sideways during the exposure, gently and repeatedly, in the hope that a flash will occur as it moves, exposing multiple strokes spread out on the film.

In the accompanying image the camera was accidentally bumped during exposure, causing the multiple strokes to expose adjacent parts of the film, revealing them in this way.

Multiple strokes seen from Erasmusrand, looking south during twilight

Daytime lightning

During daytime, unlike at night, film becomes overexposed during a time exposure lasting up to a minute. Lightning can still be photographed by releasing the shutter by hand, and will be successful if several return strokes occur.

During daylight a photographer standing ready with a finger on the shutter release may be able to capture the phase that follows after about 200 ms. This allows for a typical human reaction time of 100 ms between observing the flash and pressing the shutter release, followed by a mechanical delay in the camera and lens system of another 100 ms. Consequently the stepped leader, easily photographed at night during a time exposure, and the initial return strokes will be absent from daylight photographs.

In South Africa about half the number of flashes last 400 ms and longer, while in England for example most flashes are of shorter duration.

Daytime lightning seen from Erasmusrand, looking west, early afternoon

Strike point

The lightning channel terminates where a lightning bolt strikes the ground. The ground current, also known as the voltage gradient, is a surge of voltage flowing radially outward and dissipating through the surrounding soil. The electrical current melts sand into hollow tubular glassy objects known as fulgurites, named after fulgur, Latin for lightning, and shaped like the section of lightning creating it. Fulgurites may be up to 20 metres long and 1–5 cm in diameter.

The leftmost lightning bolt in the accompanying image shows the lightning channel continuing, unusually, towards the left on the surface. The Union Buildings are lit by floodlights, as was the custom on the first Saturday night of each month.

Direct strike seen from Sunnyside in Pretoria, looking north at night

Cloud-to-air lightning

Electrical charges build up not only in the cloud but also in the surrounding air. In cloud-to-air lightning an electrical discharge takes place between accumulated charges within a cumulonimbus cloud, usually the positively charged top, and a region of opposite charge in adjacent air. The air charges are relatively weak and usually discharged by a single flash, not producing the flickering associated by multiple strokes.

A lightning flash from the thundercloud into the upper atmosphere is sometimes called a 'sprite'. See 'Special lightning phenomena'.

The image shows a lightning flash from the top of the cloud into the upper air. In his book Basil Schonland states that such lightning is uncertain, and he probably never observed it.

Cloud-to-air lightning seen from Erasmusrand, looking south at night. Star trails may be seen, circling the southern celestial pole during the time exposure.

Intracloud lightning

Intracloud lightning is a flash within the thundercloud, from its typically negatively charged base region towards its positively charged upper levels, illuminating the cloud from within. Multiple strokes may last up to half a second, highlighting the external shape of an entire Cumulonimbus cloud.

The lightning discharge is not easily seen, being obscured by the surrounding cloud. Research by means of non-visual tools such as radio reveals that this discharge does not proceed in the familiar tortuous branching pattern of cloud-to-ground or cloud-to-cloud flashes. But where it crosses open air in between it proceeds in the branched way.

Intracloud and negative cloud-to-ground lightning are the most common types of lightning discharge. However, a very active thundercloud may sometimes produce many intracloud flashes and no cloud-to-ground flashes.

Intracloud lightning seen from Erasmusrand, looking south at night

Forked lightning

Forked lightning occurs when one or more discharge channels branch off the main discharge channel. When the lower part of the ionised channel has fizzled out for some reason, the dart leader (see 'Multiple strokes') exits into the surrounding air and the lightning proceeds in the usual stepped-leader way, finding its own way to the earth below.

The image was captured from Pretoria at night. A time exposure of about 30 s at f/8.0 was used, with a 90 mm lens. The camera was mounted on a tripod and triggered by a cable release.

Forked lightning seen from Erasmusrand, looking south-south-west towards Johannesburg, as the storm intensified and moved in

Intercloud lightning

Intercloud or cloud-to-cloud lightning is a flash between oppositely charged regions in adjacent thunderclouds, usually between the negatively charged base of one and the positively charged top of its neighbour. Intercloud lightning is less common than is often supposed. The distance between the top and base regions of adjacent clouds is greater than the typical distance between the base region of a single cloud and the ground. A greater build-up of electric charge is therefore required to overcome the thicker layer of insulating air, resulting in a much stronger flash than cloud-to-ground flashes.

Since intercloud lightning passes through clear air between clouds it is hazardous to airplanes. It may lose its way and turn downward, striking the earth 'out of the blue' kilometres away from either of the clouds that triggered it.

Since it occurs at a greater height than cloud-to-ground lightning it may be seen from further away, up to 300 km in the case of a very large Cumulonimbus cloud. Since lightning is audible up to 30 km, intercloud lightning further away will illuminate the sky without any accompanying thunder.

Lightning seemingly starting off as intercloud before turning to earth, seen from Erasmusrand

Intercloud lightning seen from Erasmusrand, looking south-west at night

The colours of lightning

Lightning is brilliantly white. The air surrounding a lightning flash may cause a shift in its colour, if it carries raindrops and other particles interacting with the light emitted by lightning. White lightning indicates low humidity in the air and is therefore most likely to cause veld fires. It is the most common kind in South Africa.

Lots of dust in the surrounding air results in yellow lightning. Blue lightning indicates the presence of hail. Green flashes are unusual. Red lightning flashes indicate rain in the surrounding air. Red may also simply indicate distant lightning, when, as with a reddish low sun or moon, the light needs to penetrate a thicker layer of atmosphere, resulting in loss of other colours because the blue end of the spectrum is scattered out by the air in-between.

Reddish lightning seen from Erasmusrand, looking west at night

Ribbon lightning seen from Erasmusrand, looking west at night

Ribbon lightning

A faint second channel may be seen to the right of the main illuminated channel. Ribbon lightning is seen when the cloud-to-ground flash channel is moved sideways by a strong wind during the component strokes of the flash. The strokes are thereby separated horizontally in space. To the eye, they appear to happen simultaneously, because of the speed of the whole sequence.

Beaded lightning seen from Erasmusrand, looking west at night

Beaded lightning

With beaded lightning the lightning channel to the ground appears to break up into luminous fragments. Various theories try to explain beaded lightning. It could be an inherent lightning phenomenon or simply a secondary effect due to perspective or atmospheric effects.

An almost necklace-like string of beaded lightning may be seen in the accompanying image. The horizontal broken line of light was created by the periodic light of an aircraft.

Sheet lightning seen from Erasmusrand, looking west at night

Sheet lightning

Sheet lightning is not a special form of lightning, but intracloud discharges that reflect into a large cloud area while the actual lightning flash is concealed behind cloud. In fact any form of lightning observed indirectly by its illumination of cloud is called sheet lightning.

In the photograph sheet lightning is seen in the white illumination of the cloud while a cloud-to-ground flash is also seen.

Stroboscopic effect

The stroboscopic effect is an artefact of time exposure during night photography. When successive lightning flashes light up a cloud and the cloud moves or grows bigger between the individual flashes, the cloud is recorded repeatedly on the same film frame. Between flashes the cloud is dark and its movement does not register on the film.

The stroboscopic effect in clouds to the west of Erasmusrand

Upward leader

When a lightning strike is imminent a mass of sparks, seen as a soft, faint glow, is sometimes seen around objects. It may assume a spherical shape. The phenomenon, seen at the top of ship's masts in earlier times, was named St Elmo's fire after the patron saint of Mediterranean sailors.

An upward leader is a stepped leader which starts at the top of a tall object such as a building and propagates upwards, sometimes triggering a lightning bolt from the thundercloud above and other times simply dissipating its charge into the air above. It may be regarded as an extreme manifestation of St Elmo's fire.

An upward leader with its origin unfortunately outside the frame. The Union Buildings of Pretoria may be seen on the horizon

Strange lightning phenomena

Ball lightning

This rare form of lightning also occurs during thunderstorms and purportedly comes in the form of a small round ball of light which may roll along the ground, move through windows and climb up objects before exploding. Doubtful researchers attribute its existence to a residual effect in the temporarily blinded eye following a flash close to the observer. A definitive photograph has never been taken. On the other hand, many anecdotal episodes are being put forward with too much consistency to be ignored and sophisticated theories have been devised to explain it.

Sprites, elves and blue jets

These phenomena occur above thunderclouds. Sprites in particular have been reported for years by pilots flying above thunderstorms. They are obscured by cloud when viewed from earth and were only scientifically investigated from flying data-gathering platforms during the last decade of the twentieth century.

Sprites and elves occur high above thunderclouds. They are assumed to be connected to positive cloud-to-ground strokes, which increase the electrical field strength above the clouds. They may be triggered by the close passage of meteors.

Sprites appear octopus-like with red heads and red, purple, or blue tentacles and occur in the upper mesosphere at altitudes of 50–90 km above the surface of the earth.

Elves appear as red disk-shaped flares which may be many kilometres wide, occurring just above sprites in the upper mesosphere and into the lower ionosphere (or thermosphere) approximately 70–100 km above the surface of the earth (see 'Weather' for the layers of the atmosphere).

Blue jets are much more rare and less is known about them. They appear as a deep blue glow flaring out of the top of thunderstorms and reaching heights of about 40 km. Unlike sprites and elves, they do not seem to be connected to low-atmosphere phenomena.

From superstition to science

Humanity's age-old fear of lightning was expressed in many cultures' religions and practices. For example, lightning is caused by a thunderbird, as evidenced by its claw-like marks on the ground and on the bark of trees (Xhosa, Basotho, North American Indians and others such as the builders of the Zimbabwe Ruins), or lightning is a snake (the San, as deduced from rock paintings).

It surfaces even today in the mildly blasphemous way the words thunder and lightning are used, more so in Afrikaans than in English. Witchdoctors used to (and still) claim to influence the weather – or explain their failure to do so by putting blame on the victims for invoking divine wrath. Unpredictable weather-related disasters in general are seen as divine intervention, for example a plane crash caused by adverse weather rather than equipment failure, or a fatality caused by lightning rather than accidental electrocution.

In Europe it was general practice for many centuries to ring church bells when a storm was imminent. The inscription Fulgura Franco, meaning 'I break up the lightning' was often inscribed on mediaeval bells. In 1786 the Parliament of Paris banned this practice on account of the many deaths of bell-ringers, 103 in a documented 33-year period of that time. In the eighteenth century the need arose to store gunpowder used for artillery. Church vaults were seen as ideal. In 1769 a church in Brescia with 100 tons of gunpowder in its vaults was struck and the explosion caused the death of 3 000 people, similar in scale to the recent Twin Towers tragedy. In 1856 about 4 000 people died in a similar disaster on the island of Rhodes.

During these times the masts of wooden ships also invited disaster, and many lightning strikes damaged, set fire to and even sank ships.

Lightning and people

A lightning stroke is lethal to human beings. There is no sure-fire protection outdoors: the only certain way to avoid being struck is to stay inside. But strokes to the ground are widespread. In a moderate lightning intensity region a home may be struck once every 100 years, or one in 100 homes struck each year. Lightning is much less life threatening than, for example, motorised transport or aggressive human beings.

When trapped outdoors in a storm with nowhere to hide, blasted by continuous flashing and thunder, one feels terribly exposed. Some comfort may be derived from not only knowing the remote statistical chance of being struck, but also realising that a storm seemingly directly overhead is more likely to be some distance away, since calculating the distance of a stroke based on the time lapse between seeing the flash and hearing the thunder is not that reliable within 8 km (see 'Distant thunder'). Most of the thunder heard nearby probably passed overhead while the flash struck the ground far away. Safety precautions further reduce the likelihood of being struck directly or by ground current.

Safety precautions outdoors

Lightning strikes in the general direction of the earth below, the stepped leader searching out the path of least resistance through the air. Only when it gets close to the ground does it seek out tall, grounded objects (see descriptions of the stepped leader and positive streamers above) and does the relative height of a human become significant. Crouching on your heels reduces your height, minimising the chances of a direct strike. Keeping your heels close together minimises the effect of the ground current generated by a nearby strike. Lying down would further reduce your height but maximise exposure to the more likely ground current. Four-legged animals are frequently killed because of the distance between their fore- and hindlegs, inviting ground currents to flow through the body.

Earthed tall objects such as isolated trees should be avoided since they are more likely to be struck and the person taking shelter underneath would be hit by ground current. People on horseback should dismount. Flat open spaces in the veld or open water should be evacuated if possible. Swimming is to be avoided since water is a good conductor of ground current. Shallow cave overhangs such as typical Drakensberg caves do not offer protection. Metallic fences may become charged not only by direct hits but also by the induced ground charge shadowing the cloud: contact, which will short-circuit the charge, should be avoided. Working with construction equipment, tractors and other farm implements is dangerous.

Strike incidents

When a person is struck by ground currents cardiac arrest is likely to occur. Bystanders able to exercise cardiopulmonary resuscitation could save lives. A direct strike is not likely to be survived, although it has been recorded. When struck directly the current flows preferentially along wet regions, such as perspiration on the skin underneath clothes and inside shoes. The heat explodes the moisture, resulting typically in clothes and shoes being blown off. Often one person in a group is killed by a direct strike while others are afflicted by the ground current emanating from the strike point and may survive. The shock wave (see 'The sound of thunder') may knock a person over and cause hearing damage. Lightning survivors may continue to suffer from psychological and various physiological scars. A specialised medical field called keraunomedicine pertains to lightning injuries.

Personal safety inside shelters

If possible, shelter should be taken when a storm approaches. For example, early warning systems at golf courses, though expensive, allow people to evacuate the exposed fields.

When lightning encounters a metal enclosure in its path the current flows around on the outside, following the easiest way. A person sitting inside a motor vehicle or aircraft is therefore reasonably protected against a direct strike when not touching the outside. The rubber tyres of a motor vehicle are of no significance in a direct hit, with the current either jumping through the air to the ground below the vehicle, flowing around the tyres, or flowing through the tyres causing them to explode.

Earthed metal objects such as plumbing or tent poles may become charged by the ground flow without having been hit directly and touching them may be fatal. Electrical cables similarly may carry ground flow, causing damage to plugged-in equipment. Telephone lines may be hit directly and talking on the telephone should be avoided. It is safer staying away from walls. Avoiding these hazards, it is quite safe inside a home. Traditional wood and mud huts, however, are unsafe.

Wildfires

With about half of all flashes current continues to flow following one of the multiple or component strokes, lasting for 1–2 tenths of a second. When exposed to the high temperature of a flash long enough, wood is set on fire. White lightning indicates low humidity in the air and is therefore most likely to cause fires. This is the typical South African situation.

Wildfire in the Organ Pipes Pass, Drakensberg

Damage to trees

Most trees struck by lightning survive. Few have their tops removed, and fewer still are completely shattered. During rain, water spirals down the tree trunk in the case of rough-barked trees. Consequently they usually suffer spiral scars when struck, only a couple of centimetres wide, with only the bark removed. With trees of relatively smooth bark, irregular patches of bark are removed when struck.

A tree possibly struck by lightning (Damaraland, Namibia)

Damage to houses

Houses may be set on fire through ignition of inflammable materials, much as with wildfires. High-resistance material can be blown apart. Installing a lightning rod is the only known way to protect a building.

The lightning rod

A lightning rod protects a conical volume downwards from its top, of ground radius equal to its height. The protection is maximal for flashes from directly above the rod but is reduced for flashes entering from the side. A lightning rod diverts lightning striking into its protected area and conducts the electrical current into the earth. It is therefore most important that a lightning rod be earthed properly. It does not avoid strikes by dissipating the ground charge. Effective dissipating systems have been developed in recent years that prevent strikes.

Lightning facts, fallacies and challenges

Scientific facts

- Thunderstorms without lightning do occur.
- Lightning may also occur in volcanic eruptions, nuclear explosions, snowstorms, sandstorms and non-thunderstorm rain.

- The electricity released by a single lightning flash would burn a household bulb for only a few months. The force of a lightning flash is derived from its intensity.
- Lightning will probably never be harnessed. Interception would be impractical. Most of its energy is lost in heating up the channel and the resulting thunder, and in light and radio waves. Furthermore, disturbing the natural balance may lead to unforeseen side effects.
- The ammonia and nitrate (fertilisers for the soil) content of precipitation is not correlated to the amount of lightning.
- Lightning may well be the force that created complex molecules found in life forms in the early development of the earth.

Popular misconceptions

- *The thunderclap is produced by the rapid expansion of heated air in the lightning channel followed by a sudden collapse of air to fill the vacuum left.* It is produced by surges in current, heating air inside the lightning channel extremely fast and high, allowing it no time to expand, which results in a supersonic shockwave which then decays into an audible acoustic wave.
- *Lightning always strikes the tallest object.* It tends to strike the tallest object around only after it has approached rather close to the ground.
- *A lightning rod discharges thunderclouds, thereby preventing lightning.* It routes lightning to the ground, should it strike in its vicinity.
- *Lightning is attracted to ironstone hills.* No scientific evidence exists that conductive geologic regions attract lightning.
- *Carrying a metal object invites lightning.* The body is already a good conductor.
- *Lightning is attracted to hot air rising from chimneys or closed bell-shaped tents.* No scientific evidence exists. During the Anglo-Boer War, many soldiers were electrocuted inside bell-shaped tents featuring a wooden metal-capped tent pole. The wooden pole acts as insulator allowing a dangerous level of electrical charge to build up in the metal cap when a storm passes overhead. A person touching the pole would shorten the insulating distance to earth and draw a spark.
- *Lightning never strikes the same place twice.* Should such safe places exist, it might have been possible to artificially construct them.
- *It is safer inside a house when the curtains are drawn, and mirrors attract lightning.* Not seeing lightning, either directly or reflected in a mirror, may be comforting but does not diminish the hazard.
- *A motor vehicle offers protection through insulation by its rubber tyres.* The metal body causes current to flow around the outside.

Scientific fallacies

Because of scant scientific data, due to major obstacles in data collection, scientific fallacies sometimes become established and repeated in books and articles.

- That intercloud (cloud-to-cloud) lightning is common is a popular misconception, according to Uman.
- Conflicting explanations and accounts of beaded lightning are presented as fact.
- Intracloud lightning is by its nature not seen directly but sometimes photographs of visible strokes are purported to illustrate it.

Scientific challenges

Many aspects of lightning remain to be unequivocally explained, for example:

- The separation and accumulation of huge charges in the build-up to lightning.
- The essence of ball lightning.
- The propagation of positive flashes.

Wind storms

Wind is one of the major forces reshaping the surface of the earth, on a time-scale ranging from geological eras during which huge masses of sand are transported, to the instantaneous small-scale reshaping of the surface by tornadoes. The wind-blown Kalahari sands form the largest continuous stretch of sand in the world, extending northward from the Gariep River through eastern Namibia and western Botswana, Angola, Zimbabwe and Zambia into the DRC. Over millions of years the Gariep River, flowing much stronger in earlier times, has carried sand acquired through erosion to the Atlantic Ocean. The north-flowing Benguela Current, itself a product of the South Atlantic anticyclonic wind system, deposits these sands on shores up the west coast of Namibia. From here it is carried inland by the dominant south-westerly winds to form the dune fields of the southern part of the Namib Desert.

The Beaufort Wind Scale

Above: Vegetation and sand-dunes near Gansbaai shaped by the wind

Wind speed may be estimated according to the Beaufort Wind Scale, by interpreting natural signs. Devised in the early nineteenth century for the Royal Navy, the original Beaufort scale indicated the type and amount of sail a ship should carry in winds of specific strengths. It was later adopted internationally for general use.

Beaufort Wind Scale			
Code	**Speed (kph)**	**Description**	**Effects on land**
0	Below 1	Calm	Smoke rises vertically
1	1–5	Light air	Smoke drifts slowly
2	6–11	Light breeze	Leaves rustle; vanes begin to move
3	12–19	Gentle breeze	Leaves and twigs move
4	20–29	Moderate breeze	Small branches move; dust blown about
5	30–39	Fresh breeze	Small trees sway
6	40–50	Strong breeze	Large branches sway; utility wires whistle
7	51–61	Near gale	Trees sway; difficult to walk against wind
8	62–74	Gale	Twigs snap off trees
9	75–87	Strong gale	Branches break; minor structural damage
10	88–102	Whole gale	Trees uprooted; significant structural damage
11	103–120	Storm	Widespread damage
12	Above 120	Hurricane	Widespread destruction

Dust devils

Dust devils occur where it is hot and dry. A rotating air mass, created by winds in the lower troposphere, combines with strong updraughts resulting from high surface temperatures to create an upward-spiralling funnel of air, carrying dust up to over 300 metres. Cumulus cloud may form over an area of powerful updraught, and is thus associated with more powerful dust devils – but normally they are harmless. Dust devils are too small in scale for their rotational direction to be influenced by the rotation of the earth, as is the case with cyclones (see also 'The climate of South Africa') and they rotate in any direction (as is the case with water swirling down a drain).

Dust devil in the Tsaugab River bed, Namibia

Dust storm in the Etosha pan, Namibia

Dust storms

By definition, a dust storm restricts visibility to less than a kilometre. It occurs where dry topsoil is created by periods of drought. Dust storms generated by powerful cold fronts (see also 'The climate of South Africa') may lift dust as high as 3 000 metres and transport it for thousands of kilometres. The dunefields of Namibia were created over aeons by dust blown inland from the coast. In the Free State, dust storms often cause serious damage by removing topsoil from farmlands.

A wet microburst near the Waterkloof military airbase in Pretoria. The upward curl of the wet microburst may be seen at the fringes near the ground

Microbursts

A microburst is a brief and powerful gust of wind radiating from a central point on the ground underneath a Cumulus congestus or Cumulonimbus cloud.

A dry microburst occurs in dry conditions when a column of rain falls onto a layer of dry air and evaporates, causing cooling, which accelerates the downdraught. The only visible sign is raised dust around the point of impact below. In a wet microburst heavy rain occurs together with evaporation. The wind and rain build up so much momentum that when the rain hits the ground it is redirected outward and upward in a distinctive curl.

Microbursts constitute a major aviation hazard.

Tornadoes

Cumulonimbus clouds sometimes have appendages protruding from the base of the cloud, as seen in the image, and are called 'mamma' clouds because of their resemblance to female mammals. These appendages are an indication that the atmosphere in the lower sections of the cloud is quite unstable and can be an indicator of impending severe weather. They form in areas of downward movement of air and may slowly vary in size. They often accompany storms but are not the cause of severe weather.

Cumulonimbus mamma over Pretoria

Tornadoes are cyclones associated with severe thunderstorms ahead of cold fronts (see 'Thunderstorms and lightning') and are therefore more common in South Africa than people generally believe. They are common over the prairies of the USA (around 750 tornadoes annually), where so-called storm chasers pursue tornadoes to witness first-hand one of nature's power displays.

Tornadoes are much smaller and shorter in duration than tropical cyclones (see below), but they are equally destructive and produce even stronger winds, exceeding 400 kph. They vary greatly in strength, lasting from a few minutes to an hour – usually about 15 minutes.

A tornado is a rotating column of air that extends from the base of a storm cloud to the ground. The path of a typical tornado is about 100 metres wide and a few kilometres long on the ground. Unless tornadoes form over inhabited areas they usually go unnoticed. The rotating wind column picks up debris along its path of destruction and the combination acts like a hole cutter moving across the land. (One of many tornado myths, still propagated in books, is that damage results from extremely low air pressure at the centre of a tornado that causes an enclosed volume of air to burst its container, for example a domestic dwelling.)

The spinning motion of a tornado is initiated by strong, high winds blowing in a different direction from winds at a lower level. As the low-pressure updraught area of the thunderstorm draws in winds, their rotation accelerates. The rotating system grows downward and emerges from the base of the cloud, from where it continues to the ground.

The Fujita Scale

The Fujita scale is used to rate the strength of a tornado. It was developed by the tornado expert Theodore Fujita of the University of Chicago.

Fujita Scale		
Scale number	**Wind speed (kph)**	**Damage type**
F0	64–117	Light
F1	118–180	Moderate
F2	181–251	Considerable
F3	252–330	Severe
F4	331–417	Devastating
F5	More than 418	Incredible

Tropical cyclones

Tropical cyclones (see 'Cyclones' under 'The climate of South Africa') originate in the tropical Indian Ocean and move in a general westerly direction, sometimes south-west into Mozambique and even into South Africa, as with Domoina in 1984. They move at up to 200 km per day and may last for weeks, covering thousands of kilometres.

Tropical cyclone 'Ando' of 2 January 2001, to the east of Madagascar. Copyright 2002 EUMETSAT

The eye of a tropical cyclone is a clear and almost calm area surrounded by bands of thunderstorm clusters spiralling outwards. They source their energy from warm (above 27° C) oceans and are seen as a mechanism to spend excess heat energy building up in the earth's atmosphere. Accumulating and dissipating huge amounts of energy, they are rather effective in this respect. They deteriorate on reaching land, when their source of energy is left behind.

Tropical cyclones share the destructive power of tornadoes (see above), but on a much bigger scale. Ranging from relatively small, with a diameter often smaller than 100 km, to huge with a diameter of up to 1 000 km, they are accompanied by sustained gale force winds of up to 300 km per hour. The accompanying intense downpours and ocean surges cause severe flooding.

Tropical cyclones cannot cross the equator since the Coriolis effect is lost there. They decay into clusters of thunderstorms on the equator. They also do not continue beyond 30 degrees latitude.

The Saffir-Simpson Scale

The National Hurricane Center in the United States uses the Saffir-Simpson scale to classify hurricanes, as tropical cyclones are called there. The scale is named after its developers, Herbert Saffir and Robert Simpson.

Saffir-Simpson Scale				
Category	**Pressure (hectopascals)**	**Wind speed (kph)**	**Storm surge (m)**	**Damage**
1	More than 980	118–152	1,2–1,6	Minimal
2	965–980	153–176	1,7–2,5	Moderate
3	945–964	177–208	2,6–3,7	Extensive
4	920–944	209–248	3,8–5,4	Extreme
5	Less than 920	More than 248	More than 5,4	Catastrophic

Wind erosion

Relatively soft sedimentary rock is slowly eroded by sand grains transported by strong winds, in a process known as aeolian erosion. (Aeolus was the Greek god of winds.) If the sandstone layer is topped by a harder layer, mushroom-shaped rocks and overhangs form. Many fine sandstone rock sculptures are to be found in the Drakensberg foothills and the Cederberg.

Wind is but one of a set of geomorphic agents and probably played a minor role in the erosion shown in the example image. Rock erosion results from a complex interplay of physical, chemical, organic and mechanical factors. Physical factors are thermal expansion and contraction. A chemical factor is calcification in certain areas, which causes a difference in resistance to weathering. Organic growth in fissures and on the surface collects water and damages the surface. Mechanical factors are wind erosion and running water.

A mushroom rock near Clarens

Dew, frost and fog

Dew, fog and cloud are formed by a process of condensation. Dew forms on the ground, fog near the ground and clouds above (see also 'Cloud Atlas').

Sublimation

Water vapour turns directly into ice or vice versa through sublimation.

Condensation

Condensation is the growth of water or ice by diffusion from contiguous water vapour. In the case of ice the term sublimation is often used. Condensation occurs when moist air reaches its dewpoint (see below) and comes into contact with condensation nuclei (see below) or a surface.

Above: Dew on grass in the morning. Notice the rainbow fragments in the out-of-focus dewdrops

Contrails (short for condensation trails), also called vapour trails, are condensation clouds that form behind aircraft, especially jets. Water vapour released by engine exhaust will condense if conditions are such that a small increase in moisture causes air to become saturated and dewpoint to be reached. Such favourable conditions occur for jet planes flying near the tropopause (see also 'Weather'). Also, exhaust particles may act as condensation nuclei for existing water vapour. At lower altitudes, aerodynamic pressure reduction behind the wings of a plane flying in warm moist air induces condensation.

Distrails (short for dissipation trails) occur when a plane flies through thin cloud and its heat and turbulence cause moisture to evaporate.

Contrail

Distrail

Condensation nuclei

The atmosphere is full of microscopic particles such as dust blown up by winds or blasted into the air by volcanic eruptions, smoke and sea salt above the oceans. These particles provide nuclei around which airborne water vapour may condense as water droplets or ice crystals.

Airborne water droplets

Tiny water droplets and ice crystals formed through condensation remain aloft through air resistance and rising air currents that overcome the pull of gravity. A water droplet has a diameter about 100 times larger than a condensation nucleus. A raindrop (see also 'Precipitation') typically has a diameter about 100 times larger than a water droplet and is therefore made up of a million tiny water droplets.

Absolute and relative humidity

Absolute humidity indicates the amount of water vapour in a given volume of air at its current temperature. Relative humidity is the ratio between what the air is currently holding and the maximum it can. In other words, saturated air has 100 per cent relative humidity.

Dewpoint

The amount of water vapour that an air mass can hold gradually decreases with cooling of the air. The dewpoint is the temperature at which air becomes saturated during cooling at constant pressure and constant water vapour content. It is followed by condensation.

For example, an air mass can hold 21,4 cubic cm of water vapour per cubic metre at 24,2 °C and only half the amount at 11,4 °C. (The dewpoint is determined by measuring the dry-bulb and wet-bulb temperatures, and using the difference as an index to a table – see also 'A private weather station', under 'Weather forecasting'.)

Dew

When the temperature of the ground or any other surface drops below the dewpoint of the air in contact with it, condensation takes place and water droplets form on the surface. Water droplets merge more readily on solid surfaces than in the air. The formation of dew is favoured by still, cloudless nights with humid air close to the ground. A deeper layer of humid air leads to the formation of fog as well, and fog does not occur without dew. The absence of a cloud blanket allows heat absorbed by the earth during the day to be lost efficiently through radiation and the surface to cool rapidly after sunset.

Spider web covered with dew

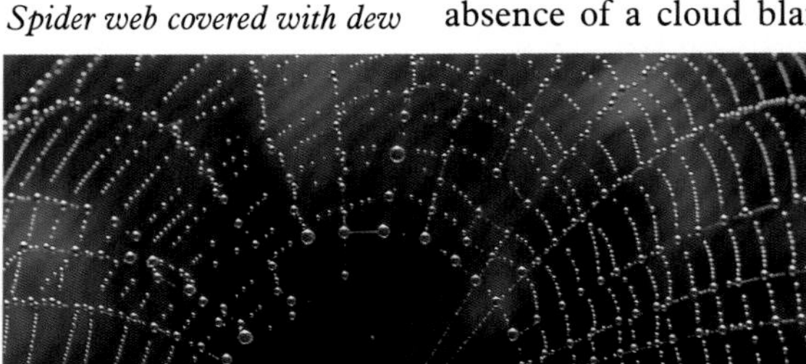

Dew does not only form in cold environments but also in hot and humid regions.

Supercooled water

Under certain conditions airborne water droplets with a temperature below the normal freezing point of water remain in liquid form and exist as supercooled water. They tend to freeze on contact with an object below freezing, for example a hailstone. Very pure water may be supercooled to –40° C, incidentally the air temperature at the top of the negatively charged region of a thundercloud (see 'Thunderstorms and lightning').

Frost

Like dew and fog, frost tends to form on clear nights when the ground temperature drops significantly through radiation – in this case below freezing. If a thin layer of moist air near the ground cools to below freezing, water vapour does not condense into dew but sublimates into ice crystals. The crystals form directly on any surface below freezing, such as stone, wood, grass and leaves, or metallic and glass surfaces. Frost crystals branch outward from the edges of whatever they form on, in beautiful intricate structures. If dew forms before the temperature falls below freezing, it freezes into solid droplets. Frost formed through sublimation is known as true frost or hoarfrost.

Frost on a glass surface

Dew drops on grass, refracting sunlight into rainbow colours

Dry air may have a dewpoint below 0° C, so moisture may not be released even when the temperature drops below freezing, and no frost will occur.

Frost presents a major weather hazard to farmers (see also 'Weather hazards').

Fog

Fog is cloud forming near the ground. In low-rainfall regions such as the Namib advection fog (see below) transports moisture from the ocean up to 50 km inland and sustains highly adapted forms of insect and plant life (see also 'Weather hazards'). Visibility is of practical concern for humans on the move and is often dangerously low for motorists and pilots. If visibility exceeds one kilometre fog is called mist.

Radiation fog

Radiation fog forms at night through radiational cooling of the earth, and the subsequent cooling of moist air in contact with the earth, in much the same way as dew (see above). It forms in addition to dew when the moist layers of air near the ground are thicker. Radiation fog is normally stationary and varies in depth from one to about 300 metres. In thick fog visibility may be reduced to three metres. Since clear skies accompany the formation of radiation fog, the following day usually has fine weather.

Advection fog

Advection is the transfer of heat by horizontal movement. It is most common at sea and near the coast. Advection fog forms when moist air moves into a cold environment or when cold air moves into a moist environment. Often similar in appearance to radiation fog, advection fog may be distinguished by its movement. Fog over sea is always advection fog since oceans never cool sufficiently to produce radiation fog.

Radiation fog at sunrise near Fouriesburg

Any fog forming during the day is likely to be advection fog. At night, moist maritime air may move inland, into an area cooled by radiation, forming advection fog. Valley fog is also most likely to be advection fog. In this case air cooled during the night becomes denser and sinks onto depressions where condensation takes place.

Advection fog at sunrise at Bloubergstrand with Table Mountain in the background

Upslope fog in the Injasuti, Drakensberg

Upslope fog

Upslope fog forms when moist air is slowly lifted to the condensation level by topography. It is most common on hills and mountains near the sea such as the Drakensberg, but may be seen in all mountain ranges. It usually begins lower down a mountain or escarpment and covers a wide area. Upslope fog differs marginally from orographic Stratus clouds (Stratus cloud lifted by topography), which result from a stronger wind and usually form near or above the peaks.

Fog stratus

After sunrise, fog formed during the night is dispersed by heat. The fog bank is eroded at the edges by direct sunlight and from underneath by the warming earth, resulting in a thinner fog bank some distance above the ground. Fog stratus forms in wind-still conditions and usually clears before noon. Thick fog stratus may produce drizzle (see 'Precipitation') or snow. Fog stratus is most common inland.

Fog stratus near Clarens, Free State

Precipitation

Precipitation occurs from a cloud when liquid droplets (rain) or ice particles (snow, graupel, hail) grow to a size and weight that can no longer be sustained by updraughts and air resistance. Ice particles may melt falling through the air and reach the ground as rain.

Warm cloud precipitation

Warm clouds are cumuliform clouds where air temperatures are above freezing and they therefore contain water droplets. Coalescence occurs as turbulence in the cloud causes water droplets to collide and merge. Wake capture occurs as small droplets are pulled into the wake of a larger drop. When drops reach a certain size and weight the forces keeping them aloft are overcome and they fall to earth as raindrops. A raindrop is about 2 mm in diameter, on average, and up to 5 mm maximum. Thinner clouds produce smaller raindrops. As raindrops fall they continue to coalesce. A rainbow often brightens towards the earth because of better refraction through bigger raindrops (see also 'Atmospheric phenomena').

Above: Highveld storm to the west of Erasmusrand

Ice crystal formation

Cold clouds are clouds with temperatures below 0° C. Supercooled water droplets exist in such a cloud at temperatures between 0° C and –40° C, while only ice crystals exist at lower temperatures. Supercooled water droplets usually coexist with ice crystals, giving rise to a mixed cloud that favours the growth of ice crystals.

Ice crystals in clouds form in various ways:

- Spontaneous nucleation occurs when tiny pure water droplets freeze at temperatures below –40° C.
- Heterogeneous nucleation occurs when droplets containing freezing nuclei freeze (see 'Condensation nuclei' above), generally at temperatures above –40° C. Within this temperature range (0° to –40° C) larger droplets freeze at correspondingly lower temperatures. Freezing nuclei insoluble in water are more efficient, and still more so if they have a crystallographic arrangement similar to ice.
- Contact nucleation occurs on contact between a freezing nucleus and a supercooled water droplet.
- Sublimation onto deposition nuclei is particularly likely when there are particles in the air.

Cold cloud precipitation

Cold cloud precipitation occurs when ice crystals grow into snowflakes in cold stratiform clouds or into hailstones in deep convective clouds.

Ice particles grow by sublimation and collision in much the same way that raindrops grow by condensation and coalescence. But the growth of ice crystals is complicated by their non-spherical shape, leading to various growth mechanisms determined by temperature. This results in the formation of thin plate, star and column snow crystals, snow flakes, graupel (soft ice pellets) and layered and lobed hailstones (hard ice pellets, see below). These various forms contribute significantly to the richness of optical atmospheric phenomena.

Virga, or precipitation from a cloud that evaporates before it reaches the ground

Rain

Rain is precipitation that reaches the ground in liquid form, whatever its origins, whether water drops or ice particles melting on their way to the ground. Smaller raindrops falling close together are called drizzle. Rain and drizzle are also categorised as light, moderate or heavy. The terms rain and shower are also used to distinguish between rain falling from stratiform clouds and Cumulus clouds respectively. Rainfall from stratiform clouds tends to be widespread and persistent, while rainfall from Cumulus clouds spawning thunderstorms may cause flash floods (see below). Red lightning flashes may indicate rain in the surrounding air or simply distant lightning (see 'Thunderstorms and lightning').

Freezing rain is common in regions that experience winter snows and therefore uncommon in South Africa. When super-cooled raindrops fall to earth they will freeze on making contact with a colder layer of air, turning rain into ice pellets in mid-air, or on impact with an object on the ground with a surface temperature below freezing. The drops spread out on impact before freezing, depositing a coating of clear ice on the object. Such precipitation is known as freezing rain or sleet.

Climatic considerations of rain are discussed under 'The climate of South Africa'.

Precipitation from a cloud that evaporates before reaching the ground is known as virga. It resembles a dark fringe hanging from the base of the cloud.

Snow

Snow begins as ice crystals (see above) that form in the middle and upper levels of the troposphere where temperatures are well below freezing. Ice crystals gradually bond into countless snowflakes, of which no two are identical. (Exactly why is explained in the book *Chaos* listed in the Bibliography.) Once heavy enough, snowflakes fall to earth.

Snow in the Long Tom Pass near Sabie

Closer to earth the growth of snowflakes is favoured by temperatures just below freezing rather than colder. This allows for the recycling of molten snow in the air, which favours the development of bigger snowflakes. Slight temperature differences swing the scale between snow and rain and it is therefore difficult to forecast snow accurately. In much colder temperatures powdery flakes form that are better for skiing.

Blizzards are caused by a combination of snow, low temperatures and strong winds. A whiteout occurs when low cloud is added and the ground and sky become indistinguishable, making navigation impossible without the use of sophisticated technology. In South Africa sporadic blizzards occur in the high Drakensberg and the Cape

Left: The Hex River Valley in April

mountains and they are dangerous to hikers and others trapped outside. In these mountains many more deaths result from exposure than from lightning, and the tragedy is that, in this case, adequate precautions are possible.

Hailstorm on the highveld

A matchbox-sized hailstone with layers of clear and rime ice

Hail

Hail forms through ice crystals growth (see 'Cold cloud precipitation' above) in the strong updraught region of thunderclouds (see 'Thunderstorms and lightning'). This region comprises a smallish part of the thundercloud and hail therefore falls in a narrow band, and not for long, maybe seven minutes, as the cloud moves along. A greenish tinge in the lower parts of a Cumulonimbus cloud or a whitening of rainfall indicate hail in a storm seen from a distance. The presence of hail may also be indicated by bluish lightning. Hail is more common in spring and summer. It is also more common in middle latitudes, since in tropical areas it melts as it falls through the warmer air to reach the ground as rain. About five hailstorms per year occur at any point on the highveld.

The strong updraughts in a Cumulonimbus cloud support hailstone formation, keeping them aloft for longer, giving them time to grow. Hail is carried into the upper ice regions of the cloud from where it falls into the supercooled layer below, in a repeated process. In warmer regions just below freezing supercooled water droplets are densely concentrated and clear layers of ice form on the hailstone. In colder parts freezing is so rapid that air bubbles are trapped and opaque rime ice forms. Alternate layers of clear and opaque ice form, more the longer the hailstone stays in the storm, but usually not more than five. Eventually gravity overcomes updraught forces and hailstones fall to earth. Hailstones are typically about 1 cm in diameter and seldom more than 3 cm. Hailstones may also agglomerate into irregular lumps. Greater hailstones cause much damage to property and crops.

Floods

Flooding of the Nile has sustained civilisation through the ages, as part of the natural annual weather cycle. Flooding of the Gariep as a result of rain upstream used to be natural before the construction of large dams. Tourists are sometimes restricted to the vicinity of their camp when the Auob and Nossob in the Kgalagadi Transfrontier Park are in flood. Irregular flooding of Sossusvlei in Namibia draws many tourists and flooding is not uncommon in deserts. Flooding sustains plant and bird life in Nylsvlei and other wetlands.

Runoff of a flash flood in the Drakensberg on the route to Cathedral Peak

In South Africa regular and extensive flooding is not utilised for the nourishment of crops and flooding is rather associated with damage, though it is not the regular major hazard that it is in other countries and continents such as Asia.

Flash flooding is caused by slow-moving thunderstorms producing concentrated precipitation that cannot be dispersed by soil absorption and runoff of water. Flash floods often occur in valleys and gorges, killing ecotourists and adventurers.

Broadscale flooding occurs with a frontal system producing prolonged and widespread rain. It builds up over days and weeks, first saturating the soil, whereafter floodwaters accumulate. Tropical cyclones cause flooding of coastal areas.

Broadscale flooding in Namaqualand

Aftermath of the 2000 flooding of the Sabie River, Kruger National Park

Drought

The smell of rain falling onto dust is characteristic of much of Africa and certainly South Africa. Drought, a natural weather hazard, is not merely low rainfall but a relative concept based on the expected rainfall for a region, whether desert or tropical, at a given time of year. The detrimental effects of drought are experienced in diverse ways by individuals and economic concerns. Various definitions are tailored to meet these diverse interests. Operational definitions help people to identify the beginning, end and severity of a drought. Conceptual definitions strive to clarify the concept of drought.

A simple drought monitoring index is rainfall measured at a specific location for a single season, expressed as a percentage of normal. Normal may be defined as average or median precipitation (the value exceeded by 50 per cent of precipitation occurrences in the long-term climate record). Many more drought monitoring indices exist.

Meteorological drought is defined on the basis of the degree of dryness in comparison to an average amount, say over 30 years, and the duration of the dry period. It is region-specific, and various definitions exist. For regions with a seasonal rainfall pattern, such as South Africa, periods of drought may be based on number of days with less precipitation than a specified value for 10-daily, monthly, seasonal or annual time scales. Hydrological drought is associated with the impact of drier periods on surface and subsurface water supply.

Agricultural drought links characteristics of meteorological and hydrological drought to agricultural impacts such as the variable susceptibility of crops during different growth stages, physical and biological properties of the soil affecting its moisture retention, and others. Socio-economic definitions of drought incorporate the supply and demand of economic goods dependent on weather-related water supply, in association with elements of the other definitions.

Normal weather may return gradually or a drought may be broken by sudden heavy rains. Droughts are one of South Africa's major weather hazards, as its many dams testify. Water is imported from Lesotho.

Auob River bed, Kgalagadi Transfrontier Park

Cloud atlas

In 1803 English scientist Luke Howard devised a classification using Latin terminology, designating three primary cloud types according to form: cirriform, stratiform and cumuliform, and their compound forms. So impressed was the German writer and poet Goethe by the simplicity and effectiveness of the system that he dedicated four poems to it. French naturalist Jean Lamarck had devised a classification of cloud forms in 1801, using French terminology, but this did not gain international recognition.

Although based on form, the classification system of Luke Howard was subsequently justified on physical grounds and therefore became accepted in meteorology The system was adopted by the International Meteorological Commission in 1929 and is used today in a modified form.

The World Meteorological Organisation publishes a standard reference volume, the *International Cloud Atlas* (see references) containing definitions as well as images of cloud types. This volume served as the primary reference for the cloud atlas in this section.

The systematic cloud classification below does not imply that clouds always comply neatly, and trained cloud observers sometimes have to revert to a category named 'chaotic sky'.

Above: Shadow rays, also known as crepuscular rays or evening rays, cast by Cumulus cloud decaying in the afternoon, seen from Erasmusrand.

Basic cloud forms

The three basic cloud forms are:

1. *Cirriform* clouds are high clouds composed of fairly widely dispersed ice crystals, resulting in relative transparency and whiteness. They do not produce precipitation.
2. *Stratiform* clouds are low clouds in uniform layers that do not exhibit individual elements. They are most common. Extended Stratus may give the sky a hazy appearance. When Stratus cloud touches the ground fog may form. In Nimbostratus form they produce precipitation, while precipitation may also result from cloud above a layer of Stratus.
3. *Cumuliform* clouds consist of individual elements, generally exhibit vertical development and are the opposite in type to the horizontal development of stratiform clouds. Convective cumuliform cloud produces heavy precipitation in Cumulonimbus form.

Cloud genera

The altitude of a point such as the base or top of a cloud is defined as the vertical distance from mean sea level to that point. The height of a point is defined as the vertical distance from the point of observation to that point.

Clouds are usually present at altitudes of up to 13 km in middle latitudes such as South Africa, up to 18 km in the tropics and up to 8 km in polar regions. For each latitude the altitude range is vertically divided into three cloud étages (levels): high, middle and low. Clouds of certain genera appear most frequently within a cloud étage. Each cloud étage is defined by an approximate height range, and the height ranges of the high and middle cloud étages overlap. The following table defines the cloud étages in terms of height and not altitude as defined above.

Cloud étages			
Cloud étage	**Height range per region (km)**		
	Polar	*Temperate*	*Tropical*
High	3–8	5–13	6–18
Middle	2–4	2–7	2–8
Low	0–2	0–2	0–2

Clouds are grouped into 10 genera: high clouds are Cirrus, Cirrocumulus and Cirrostratus, middle clouds are Altocumulus and Altostratus and low clouds are Stratocumulus and Stratus. Besides these, Cumulus clouds have low bases but develop vertically through all three levels, while Cumulonimbus and Nimbostratus are rain clouds.

Each of the cloud genera is defined at the accompanying image in the atlas. Most genera contain species and all 26 species are defined below. Various genera may have similar species; to avoid repetition, cloud species are defined below and only referred to at the applicable images.

Variations on these 36 cloud forms are defined by further descriptive terms and more than different 400 cloud types are recognised today, but not listed here. For example, see the Cumulus mamma image and description under 'Tornadoes' in the section 'Wind storms'. Orographic clouds form when mountainous terrain forces moist air blown against it to rise and condense: the 'tablecloth' of Table Mountain is a famous example. Kelvin-Helmholtz intertwined spiral cloud patterns result from strong wind shear.

Cloud species

The subdivision of cloud genera into species is based on shape and internal structure. The standard abbreviations for species are the first three letters in every case, all lower case.

1. Fibratus	Fibrous. Nearly straight or irregularly curved cloud filaments that do not terminate in hooks or tufts. Applies to Cirrus and Cirrostratus.
2. Uncinus	Hooked. Often shaped like a comma, terminating at the top in a hook, or in a tuft. Applies to Cirrus.
3. Spissatus	Thick. Cloud of sufficient optical thickness to appear greyish when viewed toward the sun. Applies to Cirrus.
4. Castellanus	Turret-like. Cumuliform protuberances in the form of turrets that appear in the upper parts of clouds. The turrets, some of which are taller than they are wide, are connected in lines. Applies to Cirrus, Cirrocumulus, Altocumulus and Stratocumulus.
5. Floccus	Woolly. Each cloud unit is a small tuft with a cumuliform appearance and ragged-looking lower part. Applies to Cirrus, Cirrocumulus and Altocumulus.
6. Stratiformis	Flattened. A cloud spread out in an extensive horizontal sheet or layer. Applies to Cirrocumulus, Altocumulus and Stratocumulus.
7. Nebulosus	Misty. A nebulous veil, showing no distinct details. Applies to Cirrostratus and Stratus.
8. Lenticularis	Lens-shaped. Cloud having the shape of an elongated lens, usually with well-defined outlines. These clouds are usually orographically induced, resulting from strong wind flow over rugged terrain. The strong flow produces a distinct wavelike pattern on the lee side of the mountain or large hill and lenticular clouds tend to form at the peaks of these waves. When very round and with well-defined edges they resemble flying saucers. Lenticular clouds are most common at the middle cloud altitudes. They are often very elongated, and occasionally show irisation (see 'Atmospheric phenomena'). Applies to Cirrocumulus, Altocumulus and Stratocumulus.
9. Fractus	Broken shape. Clouds in the form of irregular, ragged shreds. Applies to Stratus and Cumulus.
10. Humilis	Small. Clouds of limited vertical extent, usually with a flattened appearance. Base greater than height. Applies to Cumulus.
11. Mediocris	Medium-sized. Clouds of moderate vertical extent, the tops of which show protuberances. Height greater than base. Applies to Cumulus.
12. Congestus	Heaped shape. Mountainous clouds whose bulging upper parts resemble a cauliflower. Applies to Cumulus.
13. Calvus	Bald top. Cumulonimbus in which some protuberances in the upper portions of the cloud are beginning to lose their clear outlines but in which no cirriform parts can be distinguished.
14. Capillatus	Hair-like top. Cumulonimbus with distinct cirriform parts in the upper portions.

Classification tables

The following table contains the 10 genera and 26 species of cloud. These 36 cloud forms are featured in the cloud catalogue.

Cloud classification *(Heights given for middle latitudes such as South Africa)*							
Group (height, km)	**Genus (10)**	**Species (26)**					
High 5–13	Cirrus	fibratus	uncinus	spissatus	castellanus	floccus
	Cirrocumulus	castellanus	floccus	stratiformis	lenticularis	
	Cirrostratus	fibratus	nebulosus			
Middle 2–7	Altocumulus	castellanus	floccus	stratiformis	lenticularis	
	Altostratus					
	Nimbostratus					
Low 0–2	Stratus	nebulosus	fractus			
	Stratocumulus	castellanus	stratiformis	lenticularis		
Vertical development Low–10	Cumulus	fractus	humilis	mediocris	congestus	
	Cumulonimbus	calvus	capillatus			

The following table expands on the previous one, facilitating species-to-genus cross-reference.

Cloud classification *(Heights given for middle latitudes such as South Africa)*															
Group (Height, km)	Genus	Species													
		fibratus	uncinus	spissatus	castellanus	floccus	stratiformis	nebulosus	lenticularis	fractus	humilis	mediocris	congestus	calvus	capillatus
		fibrous	hooked	thick	turret-like	woolly	flattened	misty	lens-shaped	broken	small	medium	heaped	bald	hair-like
High 5– 3	Cirrus	@	@	@	@	@									
	Cirrocumulus				@	@	@		@						
	Cirrostratus	@						@							
Middle 2–7	Altocumulus				@	@	@		@						
	Altostratus														
	Nimbostratus														
Low 0–2	Stratus							@		@					
	Stratocumulus				@		@		@						
Vertical development Low–10	Cumulus									@	@	@	@		
	Cumulonimbus													@	@

Cirrus (Ci)

High, wispy clouds, usually quite thin, with a relative transparency and whiteness that allows sunlight or moonlight to pass through easily. Cirrus clouds are made up of delicate cloud filaments with fibrous (hair-like) appearance or resembling tufts or feathery plumes. They occur at temperatures lower than –25° C and are therefore composed entirely of ice crystals, usually of columnar form. They result from the widespread and prolonged ascent of air, typically by a cold front moving inland in South Africa, and indicate the probability of unsettled weather within 48 hours.

A featherlike plume of Cirrus in autumn in the sky above Pretoria

Cirrus fibratus (Ci fib)

Cirrus fibratus above weathered sandstone in the Cederberg near the Cederberg Arch. See also Cirrostratus fibratus

Cirrus uncinus (Ci unc)

Cirrus uncinus on a spring afternoon near Kamieskroon, Namaqualand. Wind shear is present, with layers of air moving at different speeds or in different directions

Cirrus spissatus (Ci spi)

Cirrus spissatus above the Sentinel, Northern Drakensberg

Cirrus castellanus (Ci cas)

A patch of Cirrus castellanus in the gap between the ridge on the left and the top of Gatberg in the Drakensberg. See also Cirrocumulus castellanus, Altocumulus castellanus and Stratocumulus castellanus

Cirrus floccus (Ci flo)

Cirrus floccus on an early autumn morning in Pretoria. See also Cirrocumulus floccus and Altocumulus floccus

Cirrocumulus (Cc)

Cirrocumulus clouds are distinct small white flakes or very small globular masses, arranged in groups, lines or delicate ripples. They are composed of ice crystals with columnar and prismatic forms and formed by widespread, slow ascent of air. The distinct patchy and/or wavelike features are common to all types of cumuliform clouds.

Cirrocumulus (higher part of the image) above the Mosterthoek Twins in the Hex River Mountains of the Western Cape

Cirrocumulus castellanus (Cc cas)

Cirrocumulus castellanus above Pretoria, distinguished by numerous small bulges in the upper parts of individual cloud elements. See also Cirrus castellanus, Altocumulus castellanus and Stratocumulus castellanus

Cirrocumulus floccus (Cc flo)

Cirrocumulus floccus above a weathered cedar tree in the Cederberg

Cirrocumulus stratiformis (Cc str)

Cirrocumulus stratiformis in a Pretoria sunset. See also Altocumulus stratiformis and Stratocumulus stratiformis

Cirrocumulus lenticularis (Cc len)

Cirrocumulus lenticularis above the Karoo, showing irisation (see also 'Atmospheric phenomena'). See also Altocumulus lenticularis and Stratocumulus lenticularis

Cirrostratus (Cs)

High clouds in the form of a whitish veil of fibrous or smooth appearance, usually blanketing the sky in ill-defined sheets that the sun or moon shines through. As with other stratiform clouds, distinct cells or sharp features usually cannot be detected. They are composed of ice crystals in the form of cubes or plates at temperatures of less than –25° C and formed by widespread, slow ascent of air. Cirrostratus sometimes creates the 22 degree halo, as in this image.

Cirrostratus in spring near Nieuwoudtville in Namaqualand

Cirrostratus fibratus (Cs fib)

Cirrostratus fibratus at sunrise in the Golden Gate National Park, with Generaalskop in the distance

Cirrostratus nebulosus (Cs neb)

Cirrostratus nebulosus and mountain zebra in the Golden Gate National Park. See also Stratus nebulosus

Altocumulus (Ac)

White or grey cloud patches of a size covered by two fingers on the outstretched arm. They are associated with the slow lifting of a large air mass and instability. The clouds are composed of flattened globules arranged in groups, rows or waves, with individual clouds sometimes so close together that their edges join. Altocumulus is sometimes difficult to distinguish from Cirrocumulus. Altocumulus clouds often produce brilliant sunsets and sunrises when lit by the sun from below.

Altocumulus at sunrise near Heuningvlei in the Cederberg, in early summer

Altocumulus castellanus (Ac cas)

Ragged tufts of Altocumulus castellanus indicate instability in the middle layers of the atmosphere above the Cederberg. These indicate instability in the middle layers of the atmosphere. See also Cirrus castellanus, Cirrocumulus castellanus and Stratocumulus castellanus

Altocumulus floccus (Ac flo)

Altocumulus stratiformis floccus in a Pretoria sunset. In non-stratiform Altocumulus floccus individual cloud patches would be more separated.

Altocumulus stratiformis (Ac str)

Altocumulus stratiformis on an early autumn afternoon in the Golden Gate National Park (the high clouds dominating the sky above the Cumulus humilis cloud on the horizon). See also Cirrocumulus stratiformis and Stratocumulus stratiformis

Altocumulus lenticularis (Ac len)

Altocumulus lenticularis around midday above the mountains near Caledon

Altostratus (As)

Greyish or bluish sheets of striated or uniform appearance, resembling thick Cirrostratus but without halo phenomena. Altostratus is typically featureless, ranging from a thin, white veil of cloud through which the sun is clearly visible as through ground glass, to a dense grey mantle that may block out the sun completely. Composed of a mixture of ice crystals and water droplets at temperatures between zero and –25° C, it is formed by widespread slow ascent of air, and is always a sign of significant amounts of moisture in the middle levels of the atmosphere.

Altostratus at sunset some distance east of Springbok in Namaqualand

Nimbostratus (Ns)

An amorphous grey cloud mass with a low and ragged base, often extending vertically well into the middle cloud region. Nimbostratus clouds are considered multi-layer clouds. Layers occur mostly in a widespread sheet. The clouds are very dark and vary in thickness. Cumulonimbus clouds are often embedded. Nimbostratus clouds are associated with large areas of continuous precipitation, which creates a diffuse appearance. They are composed of a mixture of water droplets and ice crystals, depending on the temperature, and are formed by widespread ascent of air.

Nimbostratus in summer near Wolmaransstad

Stratus (St)

Stratus clouds appear in sheets or layers. They may also be scattered, when the individual cloud elements have very ill-defined edges compared to most low cumuliform clouds (e.g. Cumulus and Stratocumulus). Fog is a low Stratus cloud in contact with the ground, which usually becomes true Stratus when it lifts. Stratus is generally grey, amorphous and uniform in shape, with a fairly uniform base. It is composed of water droplets at temperatures usually greater than –5° C and formed by the rising of moist air near the surface of the earth. It may develop into late-morning Stratocumulus, followed by early afternoon Cumulus and vertical development into Cumulonimbus.

Stratus at Pretoria

Stratus nebulosus (St neb)

Stratus nebulosus at Table Mountain. See also Cirrostratus nebulosus

Stratus fractus (St fra)

Stratus fractus above the Ndedema Valley in the Drakensberg. See also Cumulus fractus

Stratocumulus (Sc)

Grey or white cloud globules or ridges containing dark parts, which may or may not be merged. Stratocumulus clouds may be widely scattered, but are usually concentrated closer together in clusters or layers that may extend over hundreds of kilometres. Clouds usually have a ragged upper surface, with a relatively flat base and very little vertical development, lacking the sharp edges of most normal Cumulus clouds. This is the most common cloud type, formed by widespread, irregular mixing of rising air.

Stratocumulus over the quiver tree forest near Nieuwoudtville in spring

Stratocumulus castellanus (Sc cas)

Stratocumulus castellanus in a winter sunrise on the Mkaya hiking trail near Piet Retief. See also Cirrus castellanus, Cirrocumulus castellanus and Altocumulus castellanus

Stratocumulus stratiformis (Sc str)

Stratocumulus stratiformis seen from near the top of Gray's Pass in the Drakensberg. The person below the cloud cover would see Stratus clouds.

Stratocumulus lenticularis (Sc len)

Stratocumulus lenticularis, the lens-shaped cloud to the right above the mountains, seen from Cape Point. See also Cirrocumulus lenticularis and Altocumulus lenticularis

Cumulus (Cu)

Dense, detached clouds with a brilliant white sunlit portion and (usually) a grey base. Cumulus is typically puffy, with sharp outlines and noticeable vertical development in the form of towers or domes. Cells can be rather isolated or they can be grouped together in clusters. It is composed of water droplets and formed by rapid convective ascent of warm air.

Cumulus at the Kruger Gate, Kruger National Park

Cumulus fractus (Cu fra)

Cumulus fractus seen beyond Vulture's Retreat in the Drakensberg. See also Stratus fractus

Cumulus humilis (Cu hum)

Cumulus humilis in early afternoon in the Jonkershoek Valley near Stellenbosch

Cumulus mediocris (Cu med)

Cumulus mediocris at the Mooi River near Ventersdorp

Cumulus congestus (Cu con)

Cumulus congestus in a Pretoria sunset

Cumulonimbus (Cb)

Cumulonimbus are huge dense clouds in the form of a mountain with towers, usually with a dark base. They are the tallest of all clouds and may span all cloud layers, extending right through the troposphere up to the stratosphere. The temperature inversion at this height halts the rise, causing a large anvil-shaped top or plume to spread out as the vertical growth below it continues (see also 'Weather').

Cumulonimbus in the afternoon near Graaff-Reinet, above Karoo vegetation

Cumulonimbus cause most of the damage associated with severe thunderstorms, through wind gusts, lightning, hail and flash floods (see also 'Thunderstorms and lightning' and 'Tornadoes' under 'Wind storms'). The cloud is composed of water droplets in the lower part and a mixture of water droplets and ice crystals at temperatures well below 0° C in the upper part. These clouds are formed by strong convective currents with updrafts of up to 30 metres per second. If rotation occurs within the thunderstorm cell, it is called a mesocyclone.

Cumulonimbus calvus (Cb cal)

Cumulonimbus calvus on a Pretoria afternoon. Cumulonimbus or large cumulus with no anvil, although the tops may become striated

Cumulonimbus capillatus (Cb cap)

Cumulonimbus capillatus on a Pretoria afternoon

Part 2
Atmospheric Phenomena

The atmosphere consists of gas molecules, dust and pollen, water drops and droplets, ice crystals and layers of warm and cold air. When sunlight penetrates the atmosphere various optical interactions take place between the radiation and these particles. Variations are introduced by differences in the shapes and sizes of particles, and in combination they produce the visual wonders of the sky.

Light

Light is a transverse electromagnetic wave. A transverse wave vibrates in a direction perpendicular to its direction of propagation. The wavelength of visible light determines its colour. Human vision detects only a small fraction of the spectrum of electromagnetic waves, which ranges from gamma rays, x-rays and ultraviolet at the short wavelength to radio waves at the long wavelength end. In the visible spectrum violet has the shortest and red the longest wavelength. Red is followed by infrared. Forty-five per cent of the sun's radiation is in the visible range, 46 per cent in the infrared region (heat) and nine per cent in the ultraviolet region.

A ray of light naturally consists of a mixture of waves, of different wavelengths, vibrating in different planes perpendicular to their lines of propagation. The corresponding mixture of colours is integrated by the eye and sensed as another, single colour in a personal way. White light contains all visible colours, while coloured light contains only a selection. Humans are able to sense millions of subtle colours.

Light is only seen when it is reflected towards the viewer. For example, sunlight reflected from the surface of the moon is seen, but not sunlight going past.

Newton showed, by means of prisms, that white light could be broken up into its component colours, and that these colours could also cohere into white light. The sophisticated theory of light as transverse electromagnetic waves was developed by Maxwell.

Much more is known about light than what is hinted at here, while it remains mysterious in many ways. For example, light also behaves as a stream of particles or discrete quanta. This parallel theory of light is needed to explain scattering (see below), which creates the blue sky.

Reflection

The smooth surface of a mirror reflects all light falling onto it. The angle of reflection is equal to the angle of incidence and the viewer sees a virtual image of the light source behind the mirror. An object with a rough surface will reflect light in all directions, consequently no virtual image will be seen and the object itself will be seen as the light source. A water drop or an ice crystal in the atmosphere will absorb some light and reflect the rest, while more light is reflected and less absorbed with increasing angle of incidence of the light source (light coming more from the side). And these are only the beginnings of the physical events contributing to the richness of optical atmospheric phenomena ...

Polarisation

When the oscillations of all waves in a ray of light are orientated in a single plane the light is polarised. Direct sunlight is unpolarised, but reflected light is always polarised to some extent, especially so on reflection from certain surfaces such as water or foliage. Blue skylight is also polarised.

A polarising filter allows only light oscillating in one plane to pass. When looking through a polarising filter, polarised light may be extinguished through rotation of the filter. This kind of filter is often very useful in landscape and weather photography, to reduce glare that detracts from the main subject of the image. For example, it makes clouds stand out beautifully against the sky.

From a scientific point of view a polarising filter may be used to determine whether the light from an atmospheric phenomenon is polarised. If it is, reflection played a part in its creation, as for example in the rainbow where light is reflected back towards the viewer from the inside back of water drops.

Previous pages: A sun pillar in the skies above Pretoria at sunset

Refraction

Refraction is the change in direction of light as it passes obliquely into a medium with a different index of refraction, for example from water to air. The effect of such refraction is seen when a stick partially submerged in water seems to be bent sharply at the surface. The refraction index is a measure of the angle through which the direction of light changes when it is refracted.

Refraction and reflection in the atmosphere are caused by particles 10 times larger than the wavelength of the radiation. With smaller particles diffraction occurs, while even smaller particles cause scattering (see below).

Dispersion

During refraction the magnitude of the change in direction of light depends on its wavelength. White light contains light of all colours, with different wavelengths, and when it is broken down into its component colours dispersion results from these being refracted at different angles. Dispersion through water drops creates colourful phenomena such as the rainbow, with the colours always in a certain sequence. The supernumerary bows sometimes found inside the primary bow are explained by the wave theory of light and interference.

Interference

Interference is the effect produced by the superimposition of two light waves, originating from different positions, in which they alternatively reinforce or neutralise each other. Interference creates various natural phenomena such as the colours appearing in soap bubbles and iridescent clouds.

Diffraction

When light passes around the edge of a small (between a tenth and 10 times the wavelength of the light) object or hole, diffraction occurs, causing white light to break up into the colours of the spectrum and to change direction or bend. The amount of bending varies with wavelength. Effectively countless new coloured light sources come into being, over the whole surface of the hole or the circumference of the object. As the coloured light propagates in all directions the waves interfere, creating lighter and darker interference bands around the light source.

Diffraction followed by interference creates supernumerary bows inside the rainbow, the white rainbow and coronae.

Scattering

When sunlight passes through the air, the photons that make up the light beam (now the particle theory of light is applied) impact with particles in the air. This changes the direction of the light beam. Small particles, such as air molecules of radius less than a tenth of the wavelength of light, scatter shorter wavelengths markedly more than longer wavelengths. This is known as Rayleigh scattering. Therefore blue light, of shorter wavelength than the other visible colours, is scattered most, six times more than red, and the sky appears blue. Furthermore, human eyes are more sensitive to blue, making the sky appear even bluer. Distant mountains are seen through a blue haze and appear blue. Without scattering the sky would be black, as seen on photographs taken from space.

The reddish colours of the morning and evening sky result from sunlight being depleted of blue though scattering as it passes through a thicker layer of the atmosphere.

Mie scattering occurs when the radius of scattering particles and the wavelength of radiation are of similar size. The main atmospheric particles of this size are cloud droplets and pollution particles. This type of scattering is directed predominantly in the same direction as the waves. All wavelengths of visible light are affected, contributing to whiteness in the sky. (Also see 'The aureole' below.)

Physiology

Descartes, when looking through the eye of an ox with the back scraped away, saw an inverted image. This indicated the physiological nature of human sight, which inverts the upside-down image of the world falling on the retina. (Also see 'The moon illusion' below.)

Rainbow seen from the N1 to the south of Kimberley

The rainbow and its variations

Where to look

The rainbow is seen on the side of the sky that is opposite to the sun. It is a perfect circle centred on the antisolar point, which is a point on an imaginary line extending from the sun through the eye of the observer towards the centre of the rainbow. The red outside of the rainbow extends through an angle of 42° from this line and the rainbow is about 2° wide. The rainbow colour sequence from the outside is red, orange, yellow, green, blue, indigo and violet. Rainbows show up better against dark storm clouds than blue sky.

Origin

A rainbow is created when light falls onto a curtain of water drops falling through the air, whether created by rain, the spray of a waterfall or a garden hosepipe. Both sun and moon can create rainbows, although the moon rainbow is very faint, since moonlight is only about a millionth as strong as sunlight. Human night vision is devoid of colour, so that a lunar rainbow seems colourless. Its colours may however be captured on colour film. (The same applies to all but the brightest stars.)

Most light falling onto a raindrop passes through, but at the edges it is refracted and dispersed into the colours of the spectrum. It is then reflected off the inside back of the raindrop and refracted further on leaving the raindrop towards the viewer, at angles between 42° and 40° from the incoming light rays.

What is seen is the combined effect created by millions of water drops falling through the air. Each observer has his own individual circle of water drops at the right angle and antisolar point around which a rainbow forms, seen only by him. The rainbow is seen only in terms of direction and not distance. As the observer steps forward, so does his rainbow.

The reflection of light from inside water drops polarises the rainbow.

When to see

When the sun is on the horizon a complete half-circle rainbow is seen, with its top 42° high. If the sun is higher than 42°, the antisolar point will be lower by the same amount, so that the top of the rainbow would disappear below the horizon. Therefore rainbows are only seen in the morning and afternoon. Seasons and latitude influence the height of the sun, so that seasons away from summer and positions of higher latitude favour its appearance.

When viewed from an elevated post, from an aircraft or looking down at the spray of a waterfall, a rainbow may be seen at any time. From an aircraft a complete circle may be seen.

The secondary rainbow

The bright primary rainbow is sometimes accompanied by a fainter secondary bow on its outside, at an angle of 51°. Some refracted light, instead of exiting the raindrop, is reflected backwards and the process is repeated. This causes the colour sequence of the secondary rainbow to be reversed.

The sky between the primary and secondary rainbows is darker than the sky on the outside. This is known as Alexander's dark band, after the Greek sage who described it long ago. Even higher order rainbows theoretically exist and have been artificially created, but have not been seen in nature.

Variations

Rainbows may vary in colour from very colourful to the completely white fogbow. Sometimes rare supernumerary bows may be seen on the inside of a rainbow. These are featured in the following pages.

Rainbow with bright lower part

Falling raindrops merge into bigger drops on their journey through the air towards earth. Since bigger water drops produce brighter rainbows, the lower part of a rainbow near the horizon often appears more brilliant. The bigger drops are flattened in shape, falling with the circular cross-section in the horizontal plane, and are unlike the conventional teardrop shape often used by artists. In this case the rainbow is created by the circular cross-section.

North of Bethlehem on the road to Villiers, looking east at sunset

Rainbow without blue

When the sun is low in the sky, there is so much more atmosphere to be penetrated, and light towards the blue end of the spectrum gets progressively more scattered. The rainbow image here displays this effect, in that it lacks blue.

Looking east towards the University of South Africa at sunset

The fogbow

Also known as a white rainbow or mist bow, a fogbow is a pale white rainbow of low contrast and wider than the 2° of the normal rainbow. Large drops produce more vivid rainbows, while with smaller drops diffraction occurs. Wave interference dominates, causing colours to overlap and become white.

Autumn morning near Fouriesburg in the eastern Free State

Looking west into Lesotho from the Drakensberg escarpment near the Injasuti Buttress, on a summer morning

The fogbow with reddish tinge

When the water drops are slightly too large to form a completely white fogbow, a reddish tinge remains at the outside of the rainbow.

Supernumerary rainbows

Supernumerary rainbows are rarely seen. They appear just inside the primary rainbow. They are not formed in the same way as the rainbow, by refraction, but by the altogether different process of interference. The spacing of the supernumerary bows depends on drop size. Typically, falling rain contains a range of drop sizes, causing supernumerary bows to overlap and disappear.

Scene next to the N1 national road on the Worcester side of the Dutoitskloof tunnel

Crepuscular rays

Crepuscular rays or evening rays are dark shadow rays cast upon the air by clouds or mountains blocking the sun. Bright rays shine through the gaps. They appear to radiate from the sun, even when the sun is below the horizon. But they are parallel, since the sun is far away, and the appearance is due to perspective, similar to railway lines seemingly merging in the distance. They occur predominantly at sunrise and sunset but are also seen when the sun is high, and may reach hundreds of kilometres through the sky. When they are extended, they may appear to converge at the antisolar point, and are called anticrepuscular rays (see also following pages).

Seen from Erasmusrand at sunset

Seen from Erasmusrand late in the afternoon. The same phenomenon as above is seen, only with a higher sun and the rays pointing downwards to earth

Poplars in autumn at Golden Gate. The poplar trees, standing vertical, seem to converge towards their tops, illustrating the perspective effect seen with crepuscular rays

Anticrepuscular rays

Anticrepuscular rays appear to converge on the antisolar point and are extended crepuscular rays. With a low sun they have a low contrast and lovely soft purple colours.

Looking east at sunset, from Erasmusrand

The rays from a sun high in the sky shine through gaps in the cloud and apparently converge downwards and away from the viewer towards the antisolar point.

Anticrepuscular rays converging downwards. View from near the top of Gray's Pass, Drakensberg

The rainbow wheel

Anticrepuscular rays and rainbows are all centred on the antisolar point. When seen together, the anticrepuscular rays cross the rainbow perpendicular to the bow, like the spokes of a wheel. But the shadow rays on the outside of the rainbow are not seen (for reasons explained in the Bibliography). The inside rays terminate at the rainbow to create the impression of a wheel. Such rays are seen towards the left of the rainbow in the image. (This phenomenon was also painted by Constable.)

Rainbow wheel seen on a farm in the Ventersdorp district

The aureole

The aureole is the intensely bright glare around the sun (or less bright around the moon), originating from forward scattering. Scattering is the interaction and redirection of light by small atmospheric particles such as dust, pollen, smoke and tiny water droplets. Forward scattering is preferential scattering into the same direction as the light. It therefore varies in magnitude depending on conditions and does not occur when the air is extremely clear, as in high mountains. The aureole may also be seen together with coronae (see following pages}, as the white central disc inside the solar corona or lunar corona.

The pinnacle of the Inner Mnweni Needle in the Drakensberg blocking the sun

The solar corona

Solar corona in unusual cloud formation above Pretoria

A colourless solar corona seen through morning mist in the Drakensberg

Coronae are coloured concentric rings a few degrees across around the sun or moon, seen through thin clouds such as Altocumulus. The central white disc is the aureole. From there the corona colour sequence is blue, green, yellow, red. Successive rings may form with the colour sequence repeating. Since coronae are unpolarised, they do not originate from reflection. They are rather formed by the interference of light diffracted around the outside of water droplets. For a given wavelength the diffraction angle depends on the drop diameter, so that some variability in the sizes of coronae occurs. Coronae fragments may create iridescent clouds some distance away from the corona.

The lunar corona

The commoner lunar corona forms in the same way as the solar corona (see above). It is often seen near full moon as clouds float underneath the moon. The beautiful coloured rings appear when the water droplets in the cloud are of uniform diameter. Often the corona is brownish because the drop sizes vary, creating coronae of different sizes and causing the coloured rings to overlap. The bright white central disc surrounding the moon is the aureole. Favourable conditions occur during a largely clear sky when the front of a bank of cloud drifts in front of the moon.

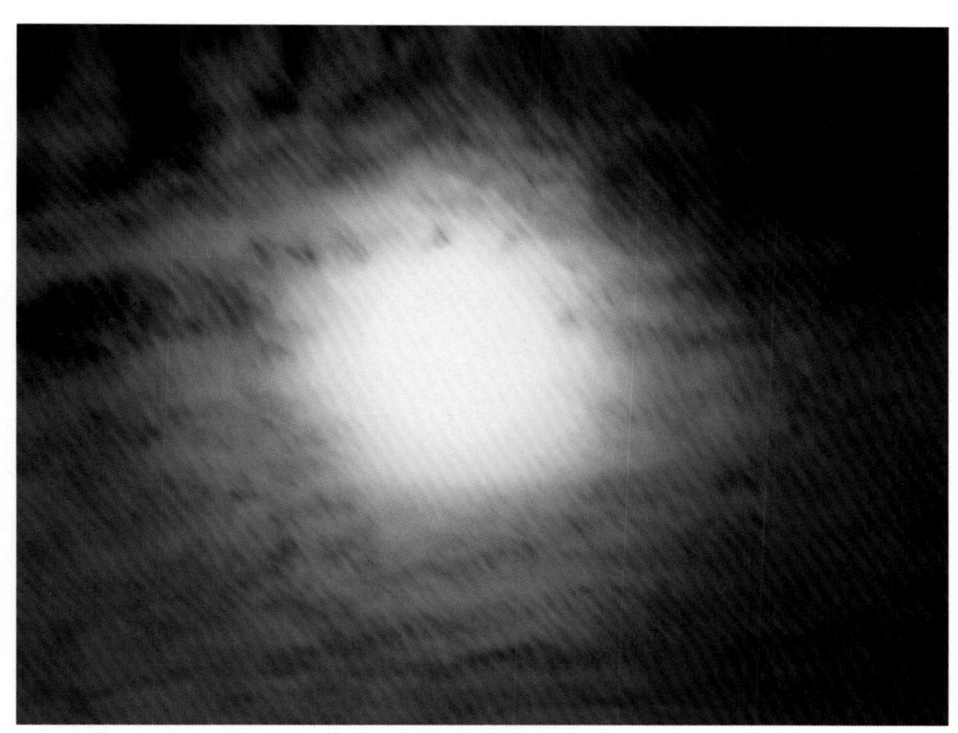

Moon and corona in the night sky above Erasmusrand

This image shows a corona formed by a crescent moon, centred not on the moon but on its brightly lit crescent. The dark part of the moon is illuminated by earthshine, sunlight reflected from the surface of the earth onto the moon.

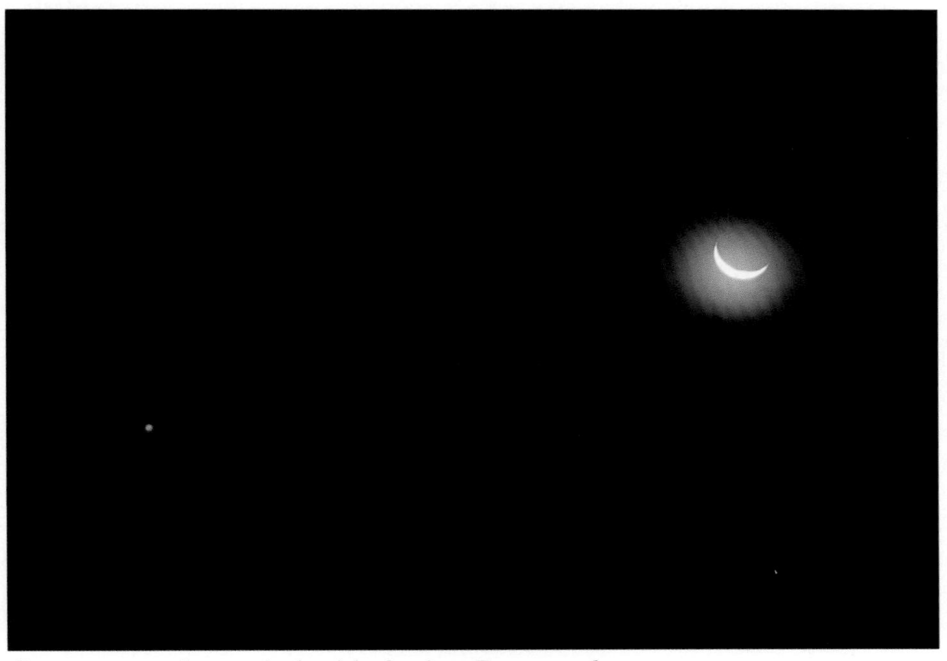

Crescent moon and corona in the night sky above Erasmusrand

The 22° solar halo

This halo forms in Cirrostratus consisting of ice crystals and appears as a white or faintly coloured silvery ring around the sun.

Haloes are created most frequently by hexagonal ice crystals, in which the most common angle of refraction is about 22°. Light is therefore refracted towards the observer at an angle of 22° away from the sun or moon. Haloes of 9°, 18° and 46° also occur. The ice crystals may be within thin layers of Cirrus cloud or free-falling. The inside of the halo is red and the outside blue. The sky inside the halo is often darker, similar to the dark band between the primary and secondary rainbow.

A solar halo seen in the skies above Erasmusrand

The partial halo seen in this image gives a better indication of its size in relation to the landscape. Looking west at sunset, from Erasmusrand.

The 22° lunar halo

Lunar haloes form in the same way as solar haloes (see above), but occur less frequently.

Halo with Jupiter on the outside seen from Erasmusrand

The circumscribed halo

The circumscribed halo appears as an elliptical halo outside the 22° halo but tangent to its upper and lower points, where it may appear as a brightening of the 22° halo. It is formed by refraction through the 60° prism of six-sided pencil-shaped ice crystals with the prism surface orientated horizontally. The shape of the halo depends on the height of the sun. Above about 30° it is smaller and more symmetrical, and collapses into the 22° halo at a solar height of about 70°. In the image the height of the sun is about 30° or slightly more.

Looking east from Erasmusrand in the morning

The mock sun

Mock suns or parhelia are found on either side of the sun at the same height as the sun. They are red on the side facing the sun and fainter and bluish away from the sun, similar to the 22° halo, but display bright colours. When the sun is low they form 22° away from the sun, but with a higher sun they are formed separately. They may display extended white tails as on this image. In the image the sun is towards the right of the mock sun and obscured by the dark cloud.

Looking west at sunset, from Erasmusrand

The image shows a close-up of a colourful mock sun seen from Erasmusrand

Sunrise and sunset colours

At sunset, or sunrise as in the image, the colours of the visible spectrum are scattered out progressively towards blue by the atmosphere, leaving the red portion. The amount of scattering depends on atmospheric haze. The glitter on the water, sometimes called the glitter path, is created by countless instantaneous reflections of the sun from waves at just the right angle to the observer. It is an optical phenomenon with only its colour influenced by the atmosphere. A sun pillar (see next) is created in a similar way when sunlight is reflected from ice crystals in the air.

Glitter path on the Gariep Dam at sunrise

A sun pillar seen in the skies above Pretoria at sunset

The sun pillar

A sun pillar is a vertical shaft of light extending above the sun, most often seen when the sun is within 1° or 2° above the horizon, as in this image. It may also be created by other light sources such as the moon and may also form below the light source. The pillar is created by the reflection of sunlight from the undersides of ice crystals in the form of plates that are slightly tipped in the vertical direction, and is therefore associated with cold air.

The countersun

The countersun is a mirage image of the sun that appears beneath the true image, at sunrise or sunset. It occurs when a thin layer of warm air lies on the surface of the earth. The warm air forms a layer of discontinuity with cooler air above, and this layer bends the rays of the sun striking it, reflecting a mirage image (see mirages later). At sunset the reflected sun rises out of the apparent horizon to join the real sun at the spot where it is about to disappear, and continues to merge with it. At sunrise the sequence is reversed.

Looking east from Twins Cave in the Drakensberg on a winter morning

Iridescent clouds

Irisation is colour in clouds from corona fragments some distance away from the sun. The coloured clouds are known as iridescent clouds. As with the occurrence of irisation in the living world, for example with insects, beautiful metallic hues may be created, indicative of overlapping corona orders and colours. Iridescent clouds may appear up to 45° away from the sun. Most favourable clouds for irisation are Altocumulus lenticularis.

Iridescent clouds seen from Erasmusrand

The beautiful metallic colours of iridescence are displayed by this image, seen from Erasmusrand

The circumzenithal arc

The circumzenithal arc is a brightly coloured arc in the form of a quarter of a circle, centred on the zenith. The zenith is an imaginary point in the sky directly above the observer. The red side of the arc faces the sun, while the blue side faces the zenith. The sun must be lower than 32° above the horizon and the arc is brightest when the sun is about 22° high, when it is found about 22° from the zenith.

Circumzenithal arc seen at Erasmusrand. The sun was excluded in the image to prevent lens glare, and is situated on the outside of the arc, just outside the field of view

The flattened sun

A flattened sun (or full moon) appearing near the horizon is caused by refraction of sunlight through the air. At lower angles the line of sight passes through a thicker layer of atmosphere and light from an object is refracted upward by a greater amount. Thus the lower parts of the sun image undergo ever more refraction. The degree of refraction depends on the altitude of the sun, the cleanness of the air and the variation of temperature through the atmosphere along the line of sight. Clean air and normal temperatures reduce the flattening. A wide range of flattening therefore occurs. It furthermore depends on the altitude of the observer, since from a higher viewpoint the horizon can be seen further away with more atmosphere in-between.

When touching the horizon the sun is geometrically already below it, but elevated through refraction as described above. Therefore slightly more than half the earth is in sunshine at any time.

Flattened sun and windsurfer at Bloubergstrand

The moon illusion

When the moon is close to the horizon it seems larger. This effect has its roots in visual physiology rather than atmospheric optics, in what is known as perceptual constancies. Distant objects are interpreted as being bigger than their angular dimensions would indicate. For example, looking at one's hand on an outstretched arm creates a smaller image on the retina of the eye, but the size of the hand is perceived to be constant whether near or far.

The sky dome is experienced as flattened, with clouds directly above being closer than those towards the horizon. Heavenly bodies are far away but also seem to be contained within this flattened dome. When the moon is seen close to the horizon, it therefore seems to be more distant and its size is interpreted as larger, creating the moon illusion. A number of other explanations are however suggested by science.

Full moon rising at Twee Rivieren, Kgalagadi Transfrontier Park

The twilight wedge

At sunset, a flat dark-blue band rises up from the eastern horizon. This is the shadow of the earth cast into space and visible in the atmosphere. It is also called the anti-twilight arch. It is best seen in clear skies, with a long line of sight. The best time is when the observer looks along the shadow's top, when sun, observer and shadow top are in the same plane, since as the shadow rises the observer looks through it at an angle. At sunrise the sequence is reversed. At the same time the twilight arch occurs in the opposite part of the sky.

The Cathedral Spur, Drakensberg, at sunrise in summer

The twilight arch

The twilight arch stretches about 90° away from the sun on either side along the horizon. Sometimes the innermost region is akin to the aureole and the solar point azimuth, the position on the horizon directly above the sun (which is below the horizon), may be detected. It is caused by sunlight scattered by the atmosphere, as the sun shines on from below the horizon. It is yellowish because sunset colours are yellowish. The lovely period of maximum contrast is short-lived and occurs when the sun is a few degrees below the horizon (say 20 minutes) after sunset or before sunrise at South African latitudes. At the same time the twilight wedge appears in the opposite part of the sky.

The twilight arch seen from Erasmusrand. The image was captured some 20 minutes after sunset during maximum contrast. The gibbous moon is included, rather small in the format because a wide-angle lens was used. This is also a good time to photograph the moon in a twilight setting, before the sky darkens more and the contrast between the sky and the moon becomes too great for the film

Some other phenomena

Many more atmospheric phenomena occur, a few of which are briefly featured here.

Lunar eclipses

During a total lunar eclipse the moon is covered completely by the shadow of the earth and should be dark. However, it is often surprisingly bright, shining in a coppery red as a result of sunlight being refracted by the earth's atmosphere onto the moon. The refracted light is red, for the same reasons that sunlight from the setting sun reaching the surface of the earth is red. The degree of coloration depends on atmospheric conditions such as cloud cover and dust from past volcanic eruptions.

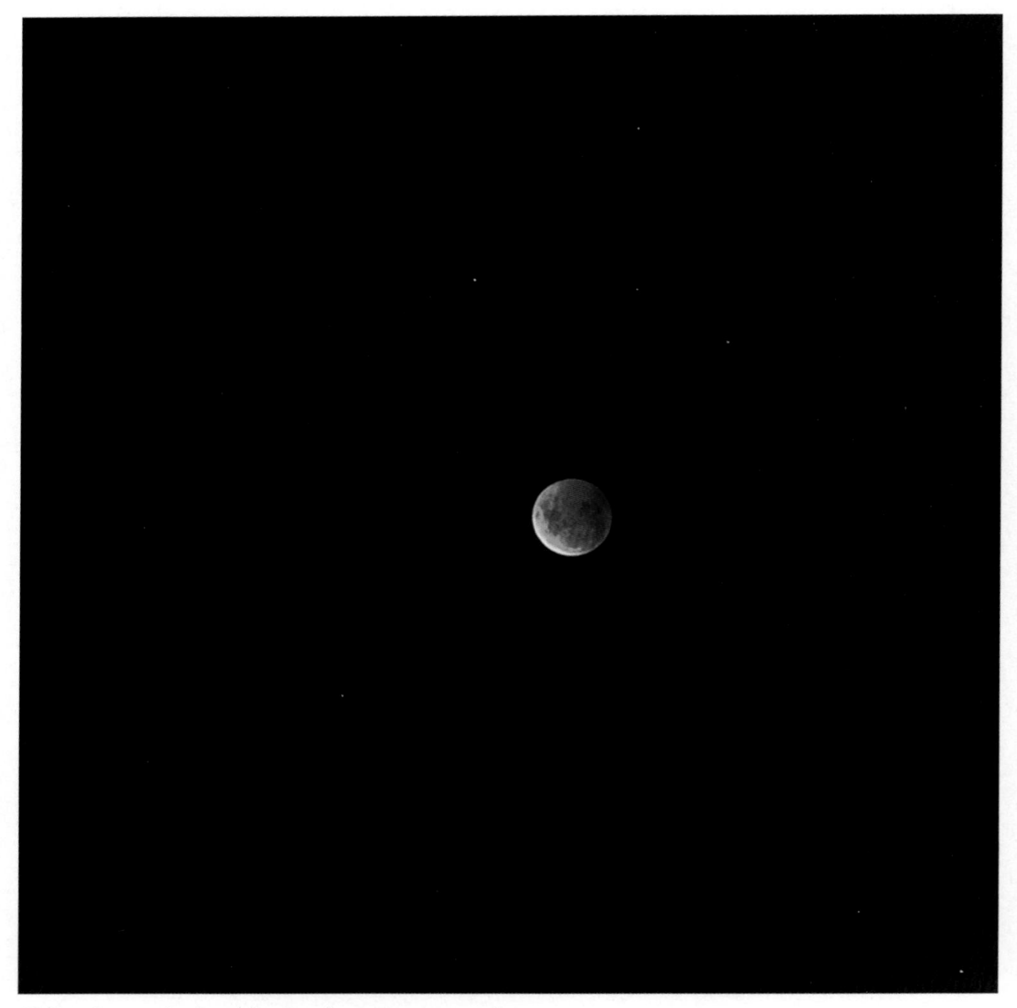

Eclipsed moon in the night sky above Erasmusrand

Eclipse shadow bands

Shadow bands are low-contrast moving bands of light on the ground, seen a few minutes before and after a total solar eclipse. They are created by layers of air of different density that refract sunlight passing through them.

Mirages

The shimmering 'water' of an atmospheric mirage seen above a desert or tarred road is blue skylight refracted upwards by a thin layer of heated air just above the surface. The temperature gradient between the heated air and air above is abnormal. (See also the countersun above.)

The green flash

The rare green flash and the even rarer blue flash occur under exceptionally clear atmospheric conditions. When the last segment of the setting sun sinks below the horizon, it appears to tear away from the sun and momentarily shines green. The phenomenon is caused by refraction near the horizon. When the sun sets its disc becomes vertically separated in overlapping images of the spectrum colours, with blue at the top followed by green and then the other colours.

If the atmosphere is not exceptionally clear the blue component is lost through scattering, and if not very clear the green component is also lost. A yellow rather than red setting sun indicates a clear atmosphere. The green flash may also be seen at sunrise if the exact position of the sun just below the horizon is known. It is more usual at higher latitudes and in clean deserts. The bright planets can also produce a green flash.

The glory

Glories are often seen from aircraft. A glory is made up of coloured circular bands on clouds, 2°–10° across, centred on the antisolar point. When seen from a plane the shadow of the plane is cast into the centre of the glory. The glory is similar to the solar corona, which is centred on the sun (see earlier pages). Unlike coronae, glories are polarised, which indicates that at least one reflection is involved, but their origin is not fully understood

The Brocken Spectre

The Brocken Spectre is the shadow of the observer, typically standing on an elevated point, cast on to mist and perceived in depth. It may be seen in combination with the glory (see above) when the shadow of the observer has a halo cast around his head.

Alpenglow

Some time after sunset, especially snow-covered mountains shine in a lovely purple that seems to brighten as the surroundings fade. The light of the setting sun, already reddish when falling onto the observer, becomes even more so as it traverses the additional band of atmosphere towards a distant mountain and back. Later the twilight arch (see above) is reflected from the mountain, which is now also seen against the backdrop of the more contrasting earth shadow (see 'The twilight wedge').

Mountain shadows

A mountain shadow as seen from the mountain always seems triangular, regardless of the shape of the mountain. As is the case with anticrepuscular rays (see above), it is a matter of perspective, with the shadow rays seemingly converging into the distance.

The twinkling of stars

The rapid changes in brightness, colour and (very slightly) position in distant point objects like stars are caused by refraction through constantly moving pockets of air of different density.

Weather in art and photography

The highly developed human faculty of visual perception is fundamental to both art and science, but, as John Constable said, 'we see nothing truly until we understand it'. In 1939, Bonacina was the first meteorologist to stress the importance of a scientific as well as artistic approach to the atmosphere.

Weather in art

John Constable was the son of a windmiller and briefly worked for his father when he was a teenager in the late 1770s. He grew up with an awareness of the vagaries of the wind and the weather behind it. He also became aware of the beauty of weather and chose to become an artist rather than follow in his father's footsteps. During those times artists were encouraged to use their imagination when painting clouds, but Constable revolted against this and became the foremost painter of realistic weather. It was at this time that Luke Howard devised his cloud classification system (see also 'Cloud atlas').

More recently, a meteorologist by the name of John E Thornes became aware of Constable's paintings and was so impressed by their realism that he investigated the artist's meteorological knowledge (see 'Bibliography') and found it to be considerable.

Above: Tranquil sunset, Erasmusrand

Constable painted clouds in a variety of forms and colours, with storms and winds of different strengths, atmospheric phenomena, atmospheric visibility and aerial perspective.

A comparable investigation into the meteorological knowledge of South Africa's painters remains to be done. The first landscape painter of stature in South Africa over a sustained period was Thomas Baines, a self-taught painter born in England in 1820. This adventurous man arrived in Grahamstown in 1848 and travelled extensively, accurately painting scenery and incidents. Jan Volschenk, born in Riversdale in 1853, became the first South African-born painter of the South African landscape. Also self-taught, he gave up his profession, accounting, at the age of 51 to devote his time to painting. Hugo Naudé, born in the Worcester district in 1869, studied in Europe and loved to paint sun-drenched landscapes. Pierneef, who painted the bushveld, was the first to shift the focus away from the Cape mountains.

Art in weather photography

The appeal of a natural landscape lies in its content, form, colour and lighting, and is often enhanced by special weather conditions. The success of a landscape photograph is determined by the extent to which the appealing features of the scene are communicated to sensitive viewers. The successful landscape photographer combines basic techniques with personal interpretation and selectiveness to evolve an individual style. Here are some guidelines:

Sandstone formation in the Cederberg: a composition with balance, geometric and aerial perspective

- Harness the experience of others, as presented in magazines, books and lectures. Join a photographic club, if only to break out of isolation and discover your mission. Viewing top quality work that touches you is a great motivator. John Thornes, mentioned above, was originally reluctant to visit an exhibition of Constable paintings – but he came away transformed. Such incidents are not unusual.

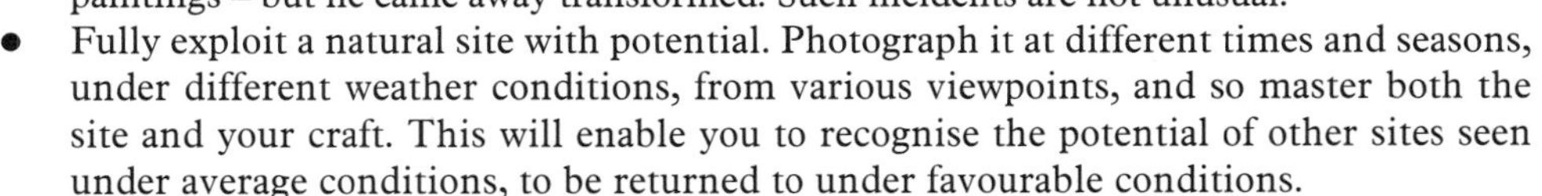

- Fully exploit a natural site with potential. Photograph it at different times and seasons, under different weather conditions, from various viewpoints, and so master both the site and your craft. This will enable you to recognise the potential of other sites seen under average conditions, to be returned to under favourable conditions.
- Create a balanced composition through the arrangement of objects that are prominent in size, colour or lighting – unless a dynamic feeling is desired and achieved through an unbalanced arrangement.
- Wide-angle lenses are effective in setting a prominent foreground against an interesting background.
- Make full use of the film format. For example, mountains photographed from a distance make disappointing images since they only occupy a narrow band on the film. Great changes may be effected simply by raising your viewpoint.
- Be aware of geometric perspective. Tilting the camera up or down changes vertical lines into converging lines and is especially disturbing with mountains, as with architecture. Stratocumulus will seem to converge when photographed along the layered lines, which may be appealing.
- Be aware of aerial perspective. A sense of depth is created when a close object is grouped with a distant object seen through some atmospheric haze.
- Utilise angled lighting, dust or mist to separate object planes. For example, a series of ridges with mist-filled valleys in between, or with the higher parts lit differently from

the lower parts, will enhance the illusion of depth beyond that of the same scene without mist or with uniform lighting, when the ridges seem to flow into each other.

- Instead of automatic exposure of unevenly lit scenes, typical under more dramatic and interesting lighting conditions, try spot or incident light metering.
- Make at least two exposures half a stop different. This enables you to select a best-exposed image and leave a backup in case of damage or loss. Seeing the effect of small adjustments will also help perfect your exposure technique. Moreover, both exposures may be valid, emphasising different aspects.

Technical notes

Equipment

All images reproduced here were captured with 35 mm equipment. Moderately wide-angle lenses are most suitable for weather images in a landscape setting. A 35 mm lens was used for most images, and 50 mm and 90 mm lenses for the remainder, with a few exceptions where long focal length lenses were used. A colourless polarising filter was frequently employed, mainly to reduce glare. No colour filters were used. Films with a wide and natural colour palette were preferred over those with a warm bias or high saturation.

Technology has a definite role to play – for the sake of both science and art – in creating images of natural integrity. Moderate colour filters may have a place in correcting colour balance distorted by a specific film under certain lighting conditions. High-contrast scenes pose serious problems to film, creating images with a distorted density range, and this is one area where digital technology will bring improvements leading to more natural images.

Atmospheric phenomena

Exposure meters are not very useful, since atmospheric phenomena typically cover a small part of the total film area and often differ greatly in brightness from the average of the scene. In the following table a summary list of atmospheric phenomena is given with exposure guidelines. Bracketing exposures by half a stop is always recommended. The exposure is given as a number of f-stops relative to the exposure for a sunlit scene with a high sun and clear weather, which is 1/125s at f/16.0 or equivalent for ISO100 film. It is also given as a shutter speed and f-stop combination. The moon, which often complements an already beautiful weather scene, is also included.

The focal length of the lens used is determined by the angular dimensions of a phenomenon and whether it is isolated or photographed in a landscape setting. To record a complete 22° halo a lens of focal length 24 mm is needed, and an even wider lens for a rainbow. A lens of focal length about 100 mm is suitable to record a lunar corona onto a sizeable portion of the film frame. A long lens of focal length f will reproduce a sun or moon image of about f/100 mm diameter on the film; for example, a 560 mm lens will record a useful 5.60 mm diameter image.

Lightning photography

Lightning photography at night is technically easy using time exposure. The difficulty lies in having a viewpoint and a storm developing within the field of view. The camera is mounted on a tripod, with the lens diaphragm set to f/8.0 for ISO100 film, and the lens is pointed at the storm. The shutter is set for time exposure and opened by means of a cable release, and the lightning flashes are recorded on the film as they occur. As soon as one or more lightning flashes have occurred the cable release is unlocked and the exposure completed. If a recording of many flashes is desired a smaller f-stop should be used.

This f-stop setting will record the faint stepped leader as well as the bright return stroke, and also distant city lights in the field of view during an exposure of up to about a minute. A much longer exposure will cause these lights to overexpose while the rest of the film will become fogged by light pollution in or near a city. It will also be fogged by lightning flashes outside the field of view reflecting off clouds inside the field of view, lightning flashes inside

Exposure guide to atmospheric phenomena		
Phenomenon	Relative brightness	Typical exposure
	f-stops	ISO100 film
Glitter path with a high bright sun; solar corona	–2	1/500s f/16.0
Solar 22° halo; mock sun	–1	1/250s f/16.0
Sky with bright white clouds	0	1/125s f/16.0
Glitter path with low sun; full moon at night	+1	1/125s f/11.0
Clouds at sunset clouds; quarter moon at night	+2	1/125s f/8.0
Crescent moon just after sunset	+3	1/125s f/5.6
	+4	1/125s f/4.0
Orange-red full moon rising in early twilight	+5	1/125s f/2.8
Twilight arch in late twilight	+6	1/125s f/2.0
Crescent moon in late twilight	+7	1/60s f/2.0
	+8	1/30s f/2.0
Lunar corona at night	+9	1/15s f/2.0
	+10	1/8s f/2.0
Earthshine on shadow part of crescent moon	+11	1/4s f/2.0
	+12	1/2s f/2.0
	+13	1s f/2.0
Fully eclipsed moon (expose < 600/f seconds)	+14	2s f/2.0
	15	4s f/2.0
Lunar 22° halo at night	+16	8s f/2.0

the field of view but obscured by clouds and shining through these clouds, and a bright moon shining onto the clouds.

A lens of 50 to 100 mm focal length is suitable. Trying to have the lightning flash fill the film frame by using a lens of longer focal length increases the likelihood of a flash occurring outside the field of view.

Twilight brings the reward of more colour in the sky and clouds. Exposure becomes more critical and a time exposure of shorter duration is necessary.

During the day or early twilight it is too bright for a time exposure. The exposure is set to the prevailing conditions and the shutter released manually as soon as the flash is observed. The stepped leader and initial return strokes will be missed because of operator and equipment reaction time delay but subsequent return strokes, should there be any, will be recorded (see also 'The dynamics of the lightning flash').

Bibliography

Major references

The following titles, detailed under the applicable subject headings below, are singled out as major reference works:

- *The Physics of Lightning*
- *Rainbows, Halos, and Glories*
- *Color and Light in Nature*
- *The Atmosphere and Weather of Southern Africa*
- *Weather – The Ultimate Guide to the Elements*
- *Chaos*

South African weather

Preston-Whyte R. A. and Tyson P. D. 1993. *The Atmosphere and Weather of Southern Africa*. Oxford University Press, Cape Town.
A comprehensive textbook on South African climate and weather.

Hurry, Lynn and van Heerden, Johan. 1983. *Suider-Afrikaanse Weerpatrone*. Via Afrika, Goodwood.
A short description of the prominent aspects of South African weather, with a catalogue of Meteosat images and corresponding weather charts.

Viljoen, Ferdie (Compiler). 1991. *Caelum – A History of Notable Weather Events in South Africa: 1500 – 1990*. Weather Bureau, Department of Environment Affairs. (Also separate updates of later events supplied by the SAWS.)
Catalogue of notable weather events in South Africa, with a few mostly monochrome photographs.

De Jager, E. J. 1999. *Drought Monitoring Products*. Abstract from presentation at the International Workshop on Optimal Use of Available Water Resources to Combat Drought and Desertification, 15-20 November 1999, Tel Aviv, Israel.
Introduction to the complexities of drought definitions and monitoring.

Bulpin, T. V. 1983. *Discovering Southern Africa*. Books of Africa, Cape Town.
Romantic travel guide by South Africa's pioneer travel guide author, with frequent references to the country's weather and climate.

Leroux, Vincent (Editor). 1992. *The Great South African Outdoors*. The Reader's Digest Association South Africa.
Voluminous nature encyclopedia with some weather and climate entries.

Symington, F. C. (Editor). 1983. *Dictionary of Meteorology and Hydrology*. Government Printer, Pretoria.
English-Afrikaans and Afrikaans-English dictionary.

The climate of South Africa

Schultze, B. R. 1994. *Climate of South Africa*. Weather Bureau, Department of Environment Affairs.
Extensive general survey of the climate of South Africa, covering sunshine, cloudiness, precipitation, evaporation, temperatures and climate regions.

Swanevelder, C. J., Kotzé, J. C. and Roos, T. J. *New Senior Geography*. Nasou, Cape Town.
Textbook with sections on South African climatology, geomorphology and oceans.

Swanevelder, C. J. and Kotzé, J. C. *Junior Geography*. Nasou, Cape Town.
Textbook with sections on South African climatology and regional geography.

Compilation of different sources. *Juta's Atlas of South Africa*. Juta, Cape Town.
Geographic atlas of South Africa including its climate regions.

General

Burroughs, William J., Crowder, Bob, Robertson, Ted, Vallier-Talbot, Eleanor and Whitaker, Richard. 1996. *Weather – The Ultimate Guide to the Elements*. Harper Collins, London.
A fine introduction and comprehensive reference volume. Compact descriptions of many weather phenomena, all of them illustrated by quality images.

Vugts, Hans F. and Beekman, F. 1997. *Wonders of Weather*. New Holland, Netherlands.
Large-format visual introduction of limited scope, with compact text and fine images.

Atkinson, B. W. and Gadd, Alan. 1986. *Weather, A Modern Guide to Forecasting*. Mitchell Beazley, London.
Introduction to weather and weather systems with an emphasis on forecasting.

Ronan, Colin A. and Dunlop, Storm. 1993. *The Skywatcher's Handbook*. The Promotional Reprint Co Ltd, for Bookmark Ltd, Leicester and Angus & Robertson, Australia.
Introduction-level guide to observation of the day and night sky.

Clouds

World Meteorological Organisation, Volume I 1975 and Volume II 1987. *International Cloud Atlas*. World Meteorological Organisation, Geneva.
The internationally recognised cloud classification system with definitions and example images.

Author unknown. 1982. *Cloud Types for Observers*. Meteorological Office, U.K.
Field guide for professional cloud observers, with colour images and descriptions.

De Jager, Elsa (compiler). 1995. *Die Wonder van Wolke*. Weather Bureau, Department of Environment Affairs and Tourism.
Information pamphlet and field guide for amateur cloud observers.

Chaos and fractal geometry

Gleick, James. 1993. *Chaos*. Abacus, UK.
Introduction to chaos theory and fractal geometry, including the role weather and clouds played in the development of the theory. Explains why snowflakes are all different and much more.

Porter, Eliot and Gleick, James. 1990. *Nature's Chaos*. Viking Penguin, New York.
Patterns in nature's disorder and wildness, as photographed by the renowned Eliot Porter. Introductory essay on chaos by James Gleick, author of the book Chaos.

McGuire, Michael. 1991. *An Eye for Fractals*. Addison-Wesley, New York.
Introductory essay on fractal patterns in nature with excellent monochrome images including a few cloud images.

Murphy, Pat and Neill, William. 1993. *By Nature's Design*. Chronicle Books, San Francisco.
Lavishly illustrated introduction to natural forms with compact explanations featuring the photography of William Neill.

Weather hazards

Goliger, A. M., Milford, R. V., Adam, B. F. and Edwards, M. 1997. *Inkanyamba, Tornadoes in South Africa*. Building Technology, CSIR and Weather Bureau.

Describes the characteristics of tornadoes and meteorological conditions favouring their development, and their annual, seasonal and geographic distribution. Contains an appendix of seven pages listing South African tornadoes from 1905 to 1999, with images of destruction but none of tornadoes.

Mills, Gus and Haagner, Clem. 1998. *Guide to the Kalahari Gemsbok National Park*. Southern Book Publishers.

An above-average tourist guide, with an emphasis on survival in a harsh climate, written by a scientist who has worked extensively in the park.

Laskin, David. 1996. *Braving the Elements*. Doubleday, New York.

A book of prose about the North American weather as it was and is experienced by its people.

Faidley, Warren. 1996. *Storm Chaser*. The Weather Channel Enterprises, Inc., Atlanta.

Lively logbook-style account of the personal experiences of a devoted storm chaser, with some great images of tornadoes and lightning.

Lane, Frank W. 1986. *The Violent Earth*. Croom Helm, Kent.

Descriptions of phenomena with historical and anecdotal notes; emphasis on weather, with relatively few images of average reproduction.

Erikson, Jon. 1988. *Violent Storms*. Tab Books, Blue Ridge Summit.

Fairly in-depth text with an emphasis on thunderstorms and lightning, hurricanes and climate change, with relatively few images of average reproduction.

Houston, Charles. 1993. *High Altitude: Illness and Wellness*. ICS Books, Merrillville, Indiana, USA.

Written by a famous mountaineer, author of *K2 the Savage Mountain*, and medical researcher into the effects of oxygen starvation on the human body.

Lightning

Malan, D. J. 1963. *The Physics of Lightning*. English Universities Press, London.

One of the first books on lightning, by Schonland's former student and later colleague giving a solid general overview; only moderately technical.

Schonland, B. 1964. *The Flight of Thunderbolts*. Clarendon Press, Oxford.

A popular book on lightning by South Africa's pioneer and in his time one of the world's leading lightning researchers, which covers history from times of superstition to the modern age of reason and science.

Austin, Brian. 2001. Schonland, *Scientist and Soldier*. Witwatersrand University Press, Johannesburg.

Extensive biography of the personal and professional life of Schonland, highlighting his lightning research at the Bernard Price Institute of Geophysical Research.

Uman, Martin A. 1986. *All About Lightning*. Dover Publications, Mineola.

A questions-and-answers-style book for the layman, answering many typical questions on the origin, variety and dangers of lightning.

Optical phenomena

Greenler, Robert. 1991. *Rainbows, Halos, and Glories*. Cambridge University Press, Cambridge.

Fine explanations, especially of ice crystal effects, with rather average image reproduction.

Lynch, David K. and Livingston, William. 2001. *Color and Light in Nature*. Cambridge University Press, Cambridge.

A recent book on this subject with many rather good images, detailed drawings and non-mathematical explanations. Contains many references to research articles at the end of each chapter.

Minnaert, M. 1954 *The Nature of Light and Color in the Open Air*. Dover Publications, New York.

A classic book and source of inspiration to many later observers and authors, which contains a few monochrome diagrams and images but detailed and informative text.

Schaaf, Fred. 1983. *Wonders of the Sky*. Dover Publications, Inc., Mineola.

A guide to observing the Earth's atmosphere, the solar system and the starry universe, written with enthusiasm, with minimal illustrations.

Murphy, Pat, Doherty, Paul and Neill, William. 1996. *The Color of Nature*. Chronicle Books, San Francisco.

Lavishly illustrated introductory book with compact explanations featuring the photography of William Neill.

Weather art and visual perception

Thornes, John E. *John Constable's Skies*. The University of Birmingham Press, Birmingham.

A meteorologist's investigation into the meteorological integrity of John Constable's paintings, with many reproductions of paintings.

Woodhouse, Bert. 1992. *The Rain and its Creatures – as the Bushmen painted them*. William Waterman Publications, a division of Ashanti Publishing (Pty) Ltd, Rivonia.

Publication of rare San rock paintings of rain, the original artistic rendering of South African weather.

Rowell, Galen. 1993. *Galen Rowell's Vision, The Art of Adventure Photography*. Sierra Club Books, San Francisco.

A collection of practical and philosophical essays by a renowned photojournalist and photographer of the natural landscape.

Levine, Michael W. and Shefner, Jeremy M. 1991. *Fundamentals of Sensation and Perception*. Brooks/Cole Publishing Company, a division of Wadsworth, Inc., California.

A textbook on the psychology of sensation and perception, readable by non-medical professionals.

Private weather station

Allaby, Michael. 1995. *How the Weather Works*. Dorling Kindersley, London.

Detailed instructions on setting up your own weather station and weather experiments, with a fine cloud gallery and clear descriptions of phenomena.

Passage of Time

Murphy, Pat, Doherty, Paul and Neill, William. 1996. *Traces of Time*. Chronicle Books, San Francisco.

Lavishly illustrated introductory book on the changing face of earth with time, often weather induced, with compact explanations and featuring the photography of William Neill.

Reimold, W. U., Brandt, D., de Jong, R. and Hancox, J. 1999. *Tswaing Meteorite Crater*. The Council for Geoscience, Geological Survey of South Africa.

Subtitle: An Introduction to the Natural and Cultural History of the Tswaing Region Including a Description of the Hiking Trail.

Index